Routledge Revivals

U.S. Timber Resources in a World Economy

This book, first published in 1967, presents a concise picture of the demand and supply trends of timber around the world. Zivnuska provides a keen analysis of plans, prospects, and opportunities in the areas covered, and an interesting look at the North American forest economy. This book will be of interest to students of environmental studies and forestry.

U.S. Timber Resources in a World Economy

John A. Zivnuska

RFF PRESS
RESOURCES FOR THE FUTURE

First published in 1967
by Resources for the Future, Inc.

This edition first published in 2015 by Routledge
2 Park Square, Milton Park, Abingdon, Oxon, OX14 4RN
and by Routledge
711 Third Avenue, New York, NY 10017

Routledge is an imprint of the Taylor & Francis Group, an informa business

Publisher's Note
The publisher has gone to great lengths to ensure the quality of this reprint but points out that some imperfections in the original copies may be apparent.

Disclaimer
The publisher has made every effort to trace copyright holders and welcomes correspondence from those they have been unable to contact.

A Library of Congress record exists under LC control number: 67021585

ISBN 13: 978-1-138-85723-0 (hbk)
ISBN 13: 978-1-315-71881-1 (ebk)

U.S. TIMBER RESOURCES IN A WORLD ECONOMY

by John A. Zivnuska

RESOURCES FOR THE FUTURE, INC.
1755 Massachusetts Avenue, N.W., Washington, D.C. 20036

Distributed by
THE JOHNS HOPKINS PRESS, *Baltimore, Maryland 21218*

RESOURCES FOR THE FUTURE, INC.

1755 Massachusetts Avenue, N.W., Washington, D.C. 20036

Resources for the Future is a non-profit corporation for research and education in the development, conservation, and use of natural resources. It was established in 1952 with the co-operation of the Ford Foundation and its activities since then have been financed by grants from that Foundation. Part of the work of Resources for the Future is carried out by its resident staff, part supported by grants to universities and other non-profit organizations. Unless otherwise stated, interpretations and conclusions in RFF publications are those of the authors; the organization takes responsibility for the selection of significant subjects for study, the competence of the researchers, and their freedom of inquiry. This book is one of RFF's resources appraisal studies, which are directed by Hans H. Landsberg. The manuscript was edited by Susan Rusinow. The illustrations were drawn by Clare O'Gorman Ford.

Director of RFF publications, Henry Jarrett; *editor,* Vera W. Dodds; *associate editor,* Nora E. Roots.

Printed in the United States of America
Library of Congress Catalogue Card Number 67-21585
Price: $3.50

PREFACE

Early conservationist thought in the United States was strongly tinged with the idea of self-sufficiency. In contemplating the future adequacy of such natural resources as metals, coal, petroleum, and timber the underlying assumption, often unspoken, was that the main problems concerned domestic supplies and domestic requirements. Small attention was paid to prospects for imports or exports.

It was from this viewpoint that Carl Schurz, when he was Secretary of the Interior nearly a century ago, warned of a timber famine, seeing only a twenty-year supply ahead. Even with all that has been learned in the meantime about what can be done through improved technology and forest management and about the significance of imports and exports, Americans have never quite lost that lingering concern.

In the early 1960's, when the staff of RFF were engaged in appraising future resources adequacy,[1] the timber outlook again appeared to have all the earmarks of eventual scarcity, unless there were enough improvements in technology, in flexibility in utilization, and in management to stem this trend. Even generous assumptions on imports did not significantly ease the prospect. The inference was not, of course, that America's forests would vanish, but rather that costs of forest products would rise and that demand would adjust by finding substitute materials.

---

[1] Hans H. Landsberg, Leonard L. Fischman, and Joseph L. Fisher, *Resources in America's Future* (Baltimore: The Johns Hopkins Press for Resources for the Future, 1963).

Even though the broad RFF study recognized the importance of the world outlook for timber, scarcity of research in that field was a handicap. It was felt at the time (and has been felt more keenly since) that among the several facets of the timber adequacy issue, the current and prospective situation outside the United States was the one that required much more analysis than it had received. More information on the world situation would throw light on the domestic outlook: Would imports relieve shortages or provide competition to U.S. producers? And how would the world situation affect U.S. export possibilities? We in RFF were happy to discover in the fall of 1964 that John Zivnuska, then about to assume the office of Dean of the School of Forestry at Berkeley, shared our interest in the world timber situation. The combination of a sabbatical leave and a grant from RFF enabled him to undertake his study in 1965.

This monograph, resulting from his work, presents a concise picture of demand and supply trends around the world; a keen analysis of plans, prospects, and opportunities in the areas covered; and an interesting look at the North American forest economy. And it provides a challenging agenda for further work.

Hans H. Landsberg
Director, Resource Appraisal Program

February, 1967

ACKNOWLEDGMENT

The dependence of any writer who attempts to survey the world scene on the knowledge and help of others is wholly obvious. In the course of this study I have drawn in the most direct terms on consultations and advice from some eighty-five individuals in seventeen nations, and in addition have drawn heavily on the writings of at least as many more. In view of the major professional assistance and frequent warm personal hospitality which I have received from so many, I am hesitant to name any in expressing my appreciation without naming all. Strong differences in viewpoint and some conflicts in data were encountered in the course of the study. The selection of the data used and the interpretations made in this report are obviously the responsibility of the writer and do not necessarily, and, in fact, could not reflect the viewpoints of all the individuals interviewed and studies consulted.

Nonetheless, I do have a special indebtedness to Dr. H. R. Josephson and his associates in the Division of Forest Economics Research of the U.S. Forest Service in this as in much of my other work. Their work has made mine possible and their cooperation has been unstinting despite occasional strong differences in our points of view. I am equally indebted in this study to those members of the Forestry and Forest Products Division of the Food and Agriculture Organization, who have carried out the remarkable series of world timber trend studies during recent years, and especially to Mr. Jack C. Westoby, Deputy Director of the Division, who keeps us all aware of the people, their needs, and their aspirations which give meaning to the timber resource. Finally, I must note my appreciation to Resources for the Future, Inc., for suggesting this study to me and supporting my work, and to the University of California for granting me the sabbatical leave during which the study was conducted.

JOHN A. ZIVNUSKA

Berkeley, California
February, 1967

TABLE OF CONTENTS

LIST OF TABLES

LIST OF FIGURES

U.S. TIMBER RESOURCES IN A WORLD ECONOMY

CHAPTER 1

INTRODUCTION

Evaluations of the timber resources of the United States have characteristically been directed to domestic timber and domestic markets. Interest has centered on the adequacy of the forests to meet present and prospective levels of consumption of forest products. Recognition has been given to foreign trade, especially imports, but only on an incidental basis in the process of evaluating sources supplemental to domestic timber in filling domestic consumption levels.

There have been good reasons for casting the problem of timber resources in these domestic terms. Logs and the forest products produced in greatest volume are heavy and bulky; costs of shipment mount rapidly and can soon become disproportionate to the values involved. Thus the forest product industries are characteristically raw-material oriented; similarly, consumption shows a strong relationship to timber availability.

Under the impact of these forces, the early American colonies faced problems of timber scarcity when settlement was still little more than a scratch on a vast forested continent. Their concern was with local timber for local use.

Although the scale has been expanded to the entire nation, the issue has continued to be analyzed primarily in such self-contained terms: domestic timber for domestic use. This emphasis has conformed to experience. From 1900 to 1950, the United States was nearly self-sufficient in timber. Net imports provided 2.5 per cent of the total wood use in the nation over this half-century.

In neither economic characteristics nor actual experience did timber and forest products appear to be commodities in which international commerce might be of compelling importance. While trade has been considered in timber resource analyses, it has understandably been relegated to a secondary position.

Signs of Change

Estimates prepared by the U.S. Forest Service [ref. 27] show that in 1904 the nation consumed the equivalent of 12.5 billion cubic feet of wood and that in 1964 total consumption was essentially unchanged, amounting to 12.4 billion cubic feet. However, in 1904 sawlogs represented 51 per cent of the total wood volume, pulpwood was 2 per cent, veneer logs were negligible, other industrial wood was 13 per cent, and fuelwood was 34 per cent. (For definition of terms, see Glossary, p. 123.) In 1964, sawlogs still represented 51 per cent of the total, but pulpwood had risen to 29 per cent and veneer logs to 9 per cent, while other industrial wood had dropped to 3 per cent and fuelwood to 8 per cent.

In very gross terms these changing percentages suggest the nature of the changes which have occurred in the forest products mix. There has been a pronounced upgrading in wood use and a rapid rise in the value added in manufacture. The low-value products such as fuelwood have declined; the increase has been

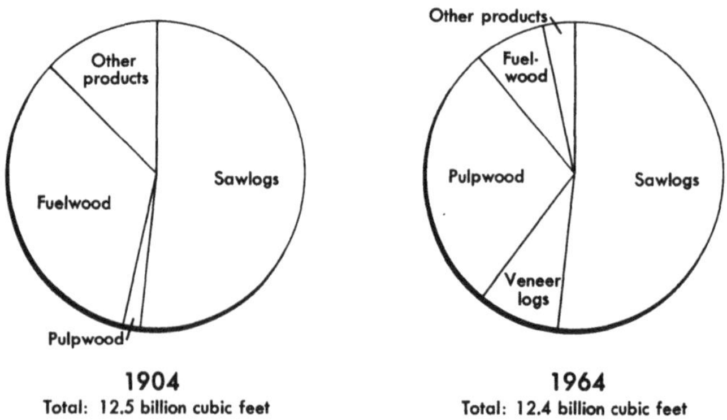

FIGURE 1. *Patterns of wood consumption in the United States, 1904 and 1964.* (Source: *U.S. Forest Service.* The Demand and Price Situation for Forest Products, 1964. *U.S. Department of Agriculture Misc. Pub. No.* 983 [*Washington: U.S. Government Printing Office, 1964*].)

in wood uses with a high degree of processing such as plywood and pulp and paper products. Even for lumber, the ratio of value to weight has increased appreciably over the years as a result of the rapid rise in prices that reached its peak in the early 1950's, and the accompanying changes in dimensions, accuracy of sawing, and extent of drying.

While forest products must still be ranked low in value-weight ratios and ability to bear high transportation costs, the product mix has been changing in a direction more favorable to trade over substantial distances.

Along with this broad pattern of change in products mix, the nation has also experienced a major change in the physical dimensions of its timber balance—the ratio of the volume of timber cut annually to the volume of the annual growth minus natural mortality. In part, the changes in ratios are apparent rather than real, reflecting changing standards of utilization and greatly improved knowledge from intensified forest surveys rather than changes in the forests themselves. However, in terms of the conception of the forest resource situation in the United States and the basis for planning, the published estimates are the valid measure of the change which has taken place over the last fifty years.

In 1920, the national cut of all timber was believed to be more than five times as high as the annual growth, the cut of sawtimber was stated to be ten times annual growth, and the exhaustion of the domestic forest resources seemed predictable [ref. 24]. This unfavorable balance was rapidly lowered in subsequent estimates, but even for 1944 the cut was shown as 2 per cent above the growth in cubic feet [ref. 25]. In contrast, the most recent estimates published by the U.S. Forest Service [ref. 28] show the annual growth in 1962 to be 60 per cent greater than the cut in cubic feet. This annual surplus of growth over cut is sufficient to increase the national inventory by nearly 1 per cent per year at current levels of growing stock.

Various imbalances by regions, species groups, and size classes exist within this over-all balance. The question has also been raised as to whether or not this rate of inventory increase is sufficient to meet the potential market opportunities of the future. Nonetheless, it must be recognized that today the United States is a timber surplus nation by a substantial margin.

Despite this current surplus of timber growth, the wood equivalent of imports into the United States in 1963 was about 3.5 times as large as the wood equivalent of the exports. On a net basis, the nation imported 13 per cent of the total consumption of forest products. In this, the United States appears to be unique. It is the only one of the economically advanced nations with a current surplus of timber growth that is not also a net exporter of forest products.

Changes, of course, have not been limited to the United States. With improved access to its extensive forests, Canada has raised its cutting rate and has become an increasingly important element in the world trade in forest products. Europe, a major consumer of forest products and traditionally a net exporter to the rest of the world, has become a net importer. During 1959–61, net imports totaled 7 per cent of wood consumption and prospects are for increased import dependency. The U.S.S.R. has regained its pre-World War II position as a major exporter of softwood lumber and has announced plans for important expansions in pulp and paper. Japan, as part of its high rate of economic growth in recent years, has become a major trading nation in forest products, importing large amounts of sawlogs and pulpwood, and exporting various manufactured wood products. And throughout the vast underdeveloped areas of the world, there is interest in two questions: developing export trade from the forest-rich areas, and obtaining needed wood products in the forest-poor areas.

These broad patterns of change can be detected in specific developments. Thus, the United States has lost its traditional pre-World War II position as a net exporter of lumber. In the last few years, lumber imports from Canada, especially from British Columbia, have rapidly increased their share of the U.S. market. The effects of a possible reduction in the existing 20 per cent tariff on softwood plywood imports have also been debated. In hardwood plywood, imports have increased from a relatively minor element in 1950 to more than half of U.S. domestic consumption by 1964. Contrary trends are not absent, however. Except for newsprint, the United States has shifted from its traditional position as a net importer of pulp and paper, becoming an increasingly aggressive net exporter. And Japanese imports of logs from the Pacific Coast have risen rapidly in volume in the last few years, placing the United States in the somewhat novel situation of a highly industrialized society exporting unprocessed raw materials which have been held to be in short supply.

Consumption is catching up with a temporary excess capacity of pulp and paper in Northern Europe; and various studies indicate that in the future, Europe will shift to appreciable import dependency in pulp and paper, as well as in lumber.

The Nature and Viewpoint of the Study

In view of the foregoing, there is need for a reassessment of the importance of present and prospective international developments to evaluations of the timber resources of the United States. This study was undertaken as a contribution to such a reassessment. It has two main objectives: First, to identify the patterns and expectations in wood availability and consumption throughout the world that may impinge on production policies and public policies affecting timber in the United States; and second, to formulate recommendations for a research program in international timber developments that can provide information useful in guiding timber production decisions and policies in the United States.

Thus, the study is intended as a problem analysis. Such a framework necessarily affects the nature of the discussion that follows. Evaluations of world patterns in timber availability and use are developed in terms of their relevance to the U.S. timber situation. From the point of view of other regions of the world, a different emphasis would be appropriate. In this sense, the study is nationalistic in orientation. Admittedly, it is not directed to placing timber in the context of the basic political and economic issue of the world in the last half of the twentieth century, which has been summarized by the United Nations Food and Agriculture Organization (FAO) [ref. 15] as "the continued concentration of the absolute growth of economic wealth in the already developed countries and the continued low level of income per capita in the countries where the greater part of the world's growth in population will take place." Nonetheless, relating U.S. timber resources to international developments may be a useful step toward understanding this broader issue.

The point of view taken regarding the nature of a resource and the role of trade are of critical importance to the analysis developed. There are various concepts of a resource; perhaps the simplest and most common view is that a resource is a stock or flow of some physical commodity or energy. From the standpoint of economic and social inquiry, however, such a concept is inadequate, since it fails to consider the questions of why and to whom such a stock or flow is a resource. There must be not only a stock or flow of some commodity or energy, but also the ability to use this stock or flow in the satisfaction of human wants and desires. Zimmerman [ref. 31] has called a resource "a function which a thing or substance may perform or . . . an operation in which it may take part, namely, the function or operation of attaining a given end such as satisfying a want." The magnitude of a resource can be seen as dependent on at least three elements: the size of the stock or rate of flow of the physical good or energy; the state of knowledge related to this stock or flow; and the nature and intensity of human wants and desires. A change in any of these three elements will change a nation's resource position.

The particular resources of concern here—U.S. domestic forest resources—satisfy human wants and desires in many different ways. Within the field of forestry, this multiplicity of functions is referred to as "multiple use," with recognition conventionally being given to such major categories as recreational and aesthetic purposes, environmental protection, grazing and browsing for domestic stock and wildlife, and timber and wood products. Most of these functions are based on the forest as an environmental factor, in which movements between nations are not possible. In contrast, the last category—timber and wood products—can be directly affected by international developments through the mechanism of foreign trade. It is this function to which the study is limited.

In considering the forest solely as a source of timber for wood products, recognition must nonetheless be given to functional distinctions not only between classes of products such as lumber, pulp, and paper, but also among grades, sizes, and other subclasses within each of these classes. Each subdivision faces its own supply problems and is affected by its own demand developments. In a functional sense, then, timber is a whole series of resources rather than a single resource. Over any given time period, timber may increase as a resource in some functions while declining as a resource in other functions. This point has been made by Duerr [ref. 4], who has noted that the same wave of progress making pulpwood resources in the South wiped out much of the firewood resource. Changes in knowledge and wants result in differential changes in timber resources related to differing functions, even though the identical physical stock of timber is considered in each function.

The question of viewpoint toward resources involves the kinds of definitional complexities which have just been noted. The view of international trade, on the other hand, is more one of attitudes and value judgments than of concept. There are a variety of these. The domestic producer understandably tends to look on imports as an undesirable source of competition and on exports as a desirable market development—but only so long as these imports and exports involve his products and not his raw materials! In contrast, the traditional conservationist, haunted by the fear of resource depletion, has tended to look on exports as an additional drain on a diminishing resource base and on imports as a helpful means of deferring the day of reckoning.

This study supports neither of these two schools. It is not concerned with the advantages of any particular commercial group, nor does it conceive of resources in a static-stock sense. The view taken is that of economic dynamics, and the concern is with economic progress. In these terms, an active international trade developed under the principle of comparative advantage appears to be advantageous to all participating nations, even though certain restrictions may sometimes be desirable for nations in the earlier stages of development.

There is no intention here to advocate any sudden development of free trade in all timber products. In an economic environment that has long been characterized by a variety of trade restraints, there are important social costs as well as social benefits in any shift toward a freer trade. Changes must be made on a deliberate and informed basis if the costs are to be held down to an acceptable level. Nonetheless, it is the viewpoint of the writer that an active international trade in timber products, both imports and exports, can enhance the values of the domestic timber resources while simultaneously contributing to freedom of choice and economic progress for the consuming public.

The report presents the findings of a single investigator working over a six-month period during the first half of 1965. Within the limits of this available time, the study has involved a literature review, travel, and interviews in seventeen nations of North America, Asia, and Europe. Time did not permit any direct studies in Africa, Latin America, or the Australia-New Zealand zone, but an effort was made to seek out and interview individuals who have had such first-hand contact.

CHAPTER 2

THE U.S. TIMBER BASE AND WOOD CONSUMPTION PATTERNS

Although the primary concern of this study is with world developments that may affect evaluations of the timber resource situation of the United States, it is desirable, first, to identify the main elements of the present situation within the United States. Therefore, this chapter will present a simple physical stocktaking of the present situation expressed primarily in terms of wood volumes.

Statistical descriptions of the timber base of the nation have been released periodically by the U.S. Forest Service for more than fifty years. The most recent such study [ref. 28] provides detailed statistics on the physical inventory as of 1962; the presentation here is based primarily on this source.

Statistics on wood consumption within the United States are surprisingly limited. Very few direct studies of wood consumption have been made on anything approaching a comprehensive basis. Instead, the estimates of wood consumption have generally been developed indirectly, relying on Bureau of the Census statistics of production, data on import and export volumes, and fragmentary data on stock changes as the basis for establishing apparent consumption. Again, various publications of the U.S. Forest Service are the best sources for comprehensive estimates of this kind.

In all of these statistics, the question of unit of measurement is a troublesome one. Each of the forest products has its own unit of measurement, developed through trade practice. Lumber is measured in board feet, veneer in square feet of surface area, softwood plywood in square feet of ⅜-inch 3-ply equivalent, hardwood plywood in square feet of surface area, poles and pilings in lineal feet, pulp and paper in tons, pulpwood in cords and various other units, and so on. Because of the historical dominance of the lumber industry in the volume of timber used by the forest product industries, the usual trade practice in the United States is to express timber volumes in board feet, although in some regions with highly active pulp industries timber volumes are also commonly expressed in cords. To add to the complexity of the situation, there are no consistent ratios for accurate conversions from one unit to another. The actual ratios vary with the size and form of the timber used.

The best common unit for expressing all these diverse units of product appears to be the cubic foot. Although the cubic foot has little trade usage in the timber and forest products field, it is a true unit of volume, independent of the shape of the product or material. It is thus a logical common measure for the wood content of the various forest products as well as for timber volumes. This has been recognized by the U.S. Forest Service, which has increasingly stressed the cubic foot as the common unit in its wood products consumption estimates and which has also provided data in cubic feet in its timber inventories. However, in its various analyses of the timber situation of the United States the Forest Service has characteristically stressed board-foot units. The justification for this is surely diminishing as the relative importance of lumber compared to other forest products declines. With the orientation of this study being toward world developments, this emphasis on the board foot is particularly unfortunate, since no comparable unit of measurement is in use outside North America. On a world basis, the cubic meter is the recognized unit of volume for expressing timber inventories and the only general unit for expressing the wood content of the various forest products. (The cubic foot is, of course, directly convertible to cubic meters at a ratio of 35.3 to 1.) In view of these considerations, primary emphasis here is placed on the expression of volumes in cubic feet. Other units are used when appropriate or necessary from the nature of the data.

The Comparability of Timber Volume Estimates

The development of current and reliable inventory data covering the forests of the nation is a complex and time-consuming operation. Although the Department of Agriculture was authorized in 1928 to conduct a national forest survey, the inventory data for 1962 are the first to be based on detailed field studies covering essentially the entire nation. Even in this instance, however, data for a number of the states are based at least in part on field studies conducted more than ten years earlier, and substantial bookkeeping adjustments have been required to bring all the material to a common date.

All earlier national data have been based only partially on completed field surveys, with estimates for major forest regions being based on incomplete field work, efforts at compilations of inventories for individual properties, and the judgment of experienced foresters, This absence of field studies has resulted in major underestimates of the U.S. timber inventory. Timber inventories also involve a long series of decisions about standards of material to be included. Over the years these standards have been changed, generally in the direction of more complete inclusion of the material actually on the ground. Thus, the successive inventory estimates are not directly comparable because of differences in the standards used in defining the material to be included.

In its two most recent analyses, the U.S. Forest Service [ref. 26, 28] has attempted to adjust the earlier estimates to comparability. Using these data, it is possible to approximate the inventory volumes at a series of past dates in terms comparable to the 1962 data and to compare these adjusted estimates with the estimates as originally published. From the vantage point of 1965, the cubic-foot volume of growing stock in 1945 appears to have been 25 per cent greater than was then reported, while the board-foot volume of sawtimber apparently was 62 per cent greater than reported.

The magnitude of the underestimations of volume in these earlier surveys deserves recognition, for this is suggestive of the uncertainties affecting many of the data used in this review of world trends. Knowledge of timber inventories in most areas of the world, including some European nations, is far less than that in the United States twenty years ago.

The Timber Base of the United States Summarized

For the purposes of the present study, only a few of the highlights of the timber inventory as of January 1, 1963, need be presented. A limited number of key statistics are summarized in Table 1.

The total commercial forest area of the United States is estimated at 509 million acres, of which 34 per cent is in the North, 40 per cent in the South, and 26 per cent in the West. The net volume of all the timber on this commercial forest area is estimated at 699 billion cubic feet. About 55 per cent of the volume is in the West, with the balance equally divided between the North and the South. Softwood species amount to 66 per cent of the total volume. The West includes 79 per cent of this softwood

TABLE 1. Selected Statistics from the United States Timber Inventory as of January 1, 1963

Item	Species group	United States	Region: North	South	West
		million acres			
Commercial forest area	All	509	172	201	136
		billion cubic feet			
Net volume of all timber[1]	Softwoods	462	34	65	363
	Hardwoods	237	122	94	21
	All	699	156	159	384
Net volume of sawtimber	Softwoods	355	16	45	294
	Hardwoods	101	50	42	9
	All	456	66	87	303
Net annual growth of growing stock, 1962	Softwoods	9.0	1.0	4.4	3.6
	Hardwoods	7.2	3.8	3.1	0.3
	All	16.3	4.8	7.5	4.0
Net annual cut of growing stock, 1962	Softwoods	7.2	0.6	2.5	4.1
	Hardwoods	3.0	1.1	1.7	0.1
	All	10.1	1.7	4.2	4.2
Surplus of growth over cut, growing stock, 1962	Softwoods	1.8	0.5	1.9	−0.5
	Hardwoods	4.3	2.7	1.4	0.3
	All	6.1	3.1	3.2	−0.3

[1] "All timber" includes sawtimber, poletimber, sound and rotten cull trees, and salvable dead trees; "growing stock" includes only sawtimber and poletimber trees.
Source: U.S. Forest Service, *Timber Trends in the United States*, Forest Resource Report No. 17 (Washington: U.S. Government Printing Office, 1965), Tables 1, 8 and 25.

volume, the South 14 per cent, and the North only 7 per cent. The volume in hardwood species is distributed very differently, with the West having only 1 per cent of the amount, the South 40 per cent, and the North 59 per cent.

Approximately 65 per cent of the total timber volume is in trees classified as being of sawtimber size. Of this sawtimber, 66 per cent is in the West, 19 per cent is in the South, and 15 per cent is in the North. The sawtimber volume has a higher proportion in softwood species than is true of timber of all sizes, with softwood species contributing 78 per cent of the total sawtimber volume. The softwood sawtimber volume is heavily concentrated in the West, which includes 83 per cent of the total. This reflects the fact that practically all of the remaining old-growth timber volume of the nation is located in the West, largely on public lands. The South has 13 per cent of the softwood sawtimber volume, while the North has only 4 per cent of the total.

Although 77 per cent of the softwood volume is in sawtimber sizes, only 43 per cent of the total hardwood volume is classified as being in sawtimber trees. The North includes 50 per cent of this hardwood sawtimber volume, the South 42 per cent, and the West 8 per cent.

The net annual growth of timber in 1962 is estimated at 16.3 billion cubic feet, or about 2.3 per cent of the total volume. In contrast, the net annual cut for 1962 is estimated at 10.1 billion cubic feet, or 1.4 per cent of the total volume. Thus the surplus of growth over cut for 1962 is 6.1 billion cubic feet, representing a surplus of growth over cut of 60 per cent and a net increase in the total timber volume of 0.9 per cent.

Regionally, this surplus of growth over cut is about equally divided between the North and the South, with the West having a slight deficit of growth as compared to cut. The surplus is heavily concentrated in hardwood species, which contributed 70 per cent of the net surplus of growth over cut in 1962, while representing only 29 per cent of the total volume of cut in that year. In contrast, western softwoods provided 41 per cent of the total cut in 1962, but growth was 0.5 billion cubic feet less than the cut.

Unfortunately, the Forest Service data on growth and cut of timber of sawtimber size are given only in board feet. With this limitation, and under the vagaries of the board-foot unit, the

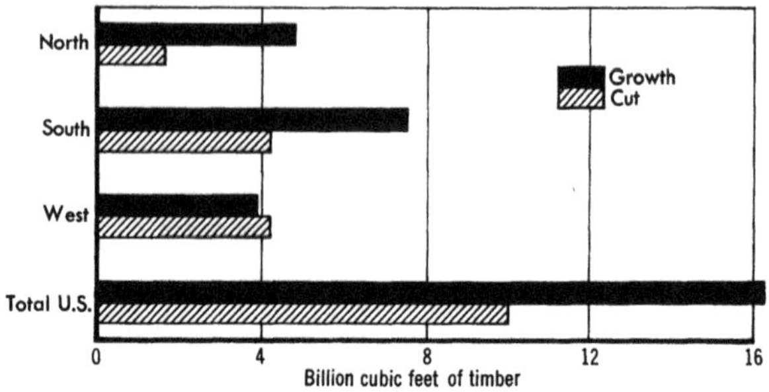

FIGURE 2. *Growth and cut of forest growing stock in the United States, 1962.* (Source: *U.S. Forest Service.* Timber Trends in the United States. *Forest Resource Report No. 17 [Washington: U.S. Government Printing Office, 1965].*)

surplus of growth over cut is relatively much less for sawtimber than for cubic-foot volume of all timber. In these special terms, the board-foot growth of all sawtimber was 13 per cent greater than the cut, but softwood growth was 2 per cent less than cut while hardwood growth was 63 per cent greater than cut.

Thus, in terms of total wood volume without regard to species, sizes, or location, the nation is currently growing substantially more timber than it is using annually. If this surplus growth, amounting to some 60 per cent of the 1962 cutting level, is to serve as a resource for the nation in a functional sense, it must be through the development of markets in which the kinds of material available can serve society. Such development could be in either domestic or world markets.

The extent to which the timber base of the nation could sustain an increase in cut is a matter of conjecture subject to many variables. Since timber is a crop, the volume of output can be directly influenced by the level of inputs devoted to timber growing. The volume of growth will also be influenced to an important degree by the cutting pattern which develops over time, including both the total amount of the cut and its distribution by regions, species, and size classes.

As a part of its most recent analysis of the nation's timber, the Forest Service attempted to project the possible development of the nation's forests under the assumption that the total cut would increase from 10.1 billion cubic feet in 1962 to 21.6 billion cubic feet in the year 2000. According to this analysis, present trends in forestry would not sustain such a more-than-doubling of the level of cut over the thirty-eight year period. A reduction of inventory and decline of growth is shown toward the end of the period. Even with this unfavorable development, a growth rate 70 per cent greater than the 1962 cutting level is indicated for the year 2000. The level of cut which could be sustained in 2000 with a somewhat less rapid expansion of cut to a position of equality with growth in that year is not shown, but it seems probable that an estimated increase in the range of 80 to 85 per cent would be consistent with the general conditions of this projection.

Obviously, the nature of the growth surpluses could not support any across-the-board expansion of the forest industries. U.S. Forest Service data indicate that only limited expansion of cutting is possible in the West, with many of these opportunities being in the Rocky Mountain region rather than in the more active producing zone of the Pacific Coast. The North could sustain a substantially larger absolute increase than the West. In view of the currently low levels of cutting in the North, this would involve cutting levels some 2½ to 3 times higher in the year 2000 than in 1962, with the cut of softwoods roughly doubling and that of hardwoods increasing threefold. By far the greatest opportunities for expansion are in the South, where roughly a doubling of the cut of both softwoods and hardwoods could be achieved.

In addition to these geographic shifts, the products mix supported by any such expansion of cutting must surely differ markedly from the products mix of 1962. Only a very limited increase can be expected in the cut of Western softwoods, which in 1962 provided 74 per cent of the nation's softwood lumber production and all of the softwood plywood production. In contrast, the increase possible in the total cut of hardwoods is roughly double the total 1962 cut, but this increase would be from smaller-sized trees and a different species mix than the 1962 cut. The development of effective markets for this material is clearly one of the great challenges and opportunities in the wood utilization field in the United States.

On the more favorable side, it deserves note that the southern pines, the nation's leading source of pulpwood and an important source of softwood lumber, had a cubic-foot growth of all timber 70 per cent greater than the cut in 1962. The current development of a softwood plywood industry in the South is one sign of response to this situation. Similarly, the cubic-foot growth of spruce and balsam fir, traditionally among the premium pulpwood species of the world, was 135 per cent greater than cut in 1962.

The foregoing comments have stressed the expansion opportunities inherent in the existing growth surpluses and current trends in forestry. It should be recognized, however, that expansion possibilities are not limited to this, since intensification of forestry efforts could expand the cutting potentials. A notable example of this is in the southern pines, where a rapid intensification of forestry practices has opened the door to possibilities for rapid expansions in output if justified by growing demands.

Furthermore, there are very important opportunities and needs in the utilization of the timber represented by the established cutting levels of 1962. Both increased utilization of the raw material in the timber cut and a further upgrading of the value of the products now produced can be significant means for increasing the role of timber in the resource base of the nation.

Thus, the possible relationship of world developments to the timber resource of the United States must be approached in terms of the present physical basis of this resource in the nation, but with full recognition of the dynamics of the situation.

Patterns of Wood Consumption in the United States[1]

Conventionally, estimates of fuelwood consumption are separated from estimates of the industrial use of wood in lumber manufacturing, pulp and paper production, plywood, and similar forms. Combining both fuelwood and industrial wood, the total consumption of wood in the nation apparently reached its all-time peak in 1907. Consumption then declined gradually, with a rapid drop during the Great Depression to a low in 1932, and then rose again to post-World War II peaks in 1956 and 1964.

[1] As noted earlier, estimates of the consumption of forest products in the United States are usually derived from Bureau of the Census reports on production, with adjustments for imports and exports, and for the limited data on changes in stocks. Various special studies and trade association reports also contribute to these estimates. Such estimates are available, for example, from the Division of Forest Economics and Marketing Research of the U.S. Forest Service, which has been particularly active in compiling material on forest products consumption. The data on past and current consumption levels included in this discussion are drawn primarily from one of the annual reports of the Forest Service [ref. 27].

Considering the tremendous changes which have taken place in the general economy of the nation over this period, annual wood consumption for all purposes in the United States during the twentieth century can only be considered as strikingly stable in total quantity.

In the early part of this period, fuelwood represented an important part of the total, amounting to 40 per cent in 1900. Fuelwood consumption gradually declined in volume until 1929, rose sharply to a depression-related peak in 1933, and then began to decline again. By 1964, fuelwood consumption was only 8 per cent of the total wood use. In view of this minor and declining importance of fuelwood and the fact that its characteristics are such that it rarely enters into foreign trade, this study concen-

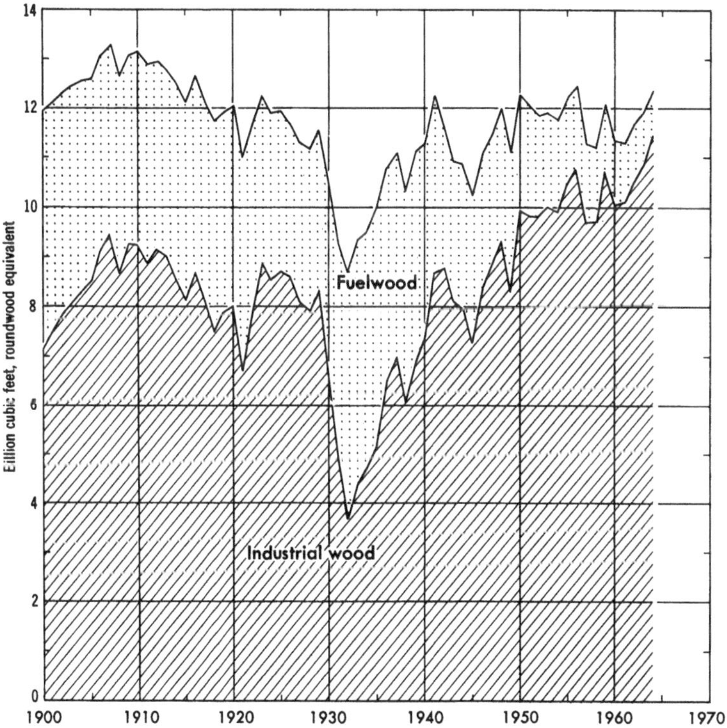

FIGURE 3. *Apparent consumption of timber products in the United States, 1900–64.* (Source: *U.S. Forest Service.* The Demand and Price Situation for Forest Products, 1964. *U.S. Department of Agriculture Misc. Pub. No.* 983 [*Washington: U.S. Government Printing Office, 1964*].)

trates on industrial uses of wood, and considers fuelwood only incidentally.

The industrial use of wood in the United States rose rapidly to a peak of 9.4 billion cubic feet in 1907, which was related to the very high level of lumber consumption of that period. The total then declined gradually until 1929, with the depression bringing a sharp drop to a low of 3.7 billion cubic feet in 1932. Consumption then began to rise again, but it was not until 1948 that the peak level of 1907 was equaled. During the 1950's, total consumption fluctuated between 9.8 and 10.8 billion cubic feet, while 1964 brought the highest level estimated to that time, amounting to 11.4 billion cubic feet.

Of this apparent consumption of 11.4 billion cubic feet in 1964, sawlogs amounted to 6.4 billion cubic feet, pulpwood to 3.5 billion cubic feet, veneer logs to 1.1 billion cubic feet, and other products to 0.4 billion cubic feet. Thus, lumber remains the largest element in the industrial use of wood, amounting to 56 per cent of the total. However, the relative importance of lumber has declined over the years, while pulpwood and veneer logs have risen in importance.

The consumption of lumber increased at a rapid and accelerating rate during the last half of the nineteenth century, reaching its peak level of around 43 billion board feet in the period 1905–10. Per capita consumption in 1905 is estimated at 503 board feet. Total consumption then began to decline, dropping to 37 billion board feet in 1929. Following a disastrous depression low of 16 billion board feet, consumption recovered to 28 billion board feet in 1939. After World War II, consumption rose to 38 billion feet in 1948 and to peaks of 41 billion board feet in 1950 and 1956. In 1964, consumption apparently totaled 41.3 billion board feet, giving a per capita consumption of 215 board feet, which is well below the level of the early 1950's. The decline from the 1905–10 peak period is explained by a number of factors, including the ending of the period of agricultural expansion, increasing urbanization, the development of alternate materials, and the very rapid rise of lumber prices compared with prices generally. However, it must be noted that lumber prices reached their peak in 1950, and subsequently have tended to move down slightly compared with prices generally. Thus, while consumption in 1964 was apparently 1 per cent higher than in 1950 or 1956, this was at price levels about 7 per cent lower in constant dollar terms. Although supply shortages may have been important in checking lumber consumption in an earlier period, there can be little doubt that over the last ten or fifteen years the demand for lumber has sagged despite great increases in over-all economic activity. The demand for one-inch boards has been particularly weak, with the demand for structural sizes being somewhat stronger.

In 1963, softwoods represented 82 per cent of the apparent consumption of lumber; hardwoods provided the remaining 18 per cent. With some minor variations, this ratio of four-fifths softwoods to one-fifth hardwoods in lumber has persisted for many years.

The pronounced rise in the importance of pulpwood reflects the rapid rise in the consumption of paper and paperboard in the United States. In 1962, total consumption of paper and paperboard amounted to 42.4 million tons; this was 2½ times the 1940 level, and more than 5 times the 1920 level. In addition to the wood pulp used in manufacturing paper and paperboard, about 1.1 million tons of dissolving pulp were consumed domestically in 1962 in the manufacture of rayon, cellophane, plastics, and similar products.

Reflecting this rapid rise in paper and paperboard consumption, the apparent consumption of pulpwood in all such products used in the United States has risen steadily from 0.4 billion cubic feet in 1910 to 3.5 billion cubic feet in 1964. The various sources of this material have shifted in relative importance over the years. For example, from 1954 to 1964, domestic production of pulpwood increased from 1.9 billion cubic feet to 2.7 billion cubic feet, while the pulpwood equivalent of all net imports of pulpwood, pulp, and paper was stable over the period at slightly less than 0.8 billion cubic feet. Considering all sources, hardwoods supplied about 34 per cent of the increase in pulpwood consumption from 1954 to 1964, with their share of the total pulpwood equivalent consumed rising from 12 per cent in 1954 to 19 per cent in 1964.

The equivalent volume of veneer logs represented by veneer and plywood consumption in the United States has shown a particularly rapid relative increase. This volume has risen steadily from 0.08 billion cubic feet in 1920 to 1.1 billion cubic feet in 1964. The major part of this volume has been in softwoods.

Other industrial wood products include a wide range of commodities such as cooperage, poles and piling, fence posts, hewn ties, shingle bolts, and excelsior bolts. Consumption of wood for this group of products reached a peak of about 2.0 billion cubic feet annually during the period from 1907 to 1912, and then began to decline, amounting to only 0.4 billion cubic feet in 1964.

The comparison below of Forest Service estimates for 1925 and 1962 shows that total industrial wood consumption in the United States has risen since the middle 1920's as a result of a rapid growth in the pulp and paper industry and the plywood industry, which has more than offset a slight decline in lumber consumption and a sharp decrease in other wood products.

Product	1925	1962	Change, 1925 to 1962
	Billion cubic feet, roundwood equivalent		
Sawlogs	6.3	5.9	−0.4
Veneer logs	0.1	1.0	+0.9
Pulpwood	0.8	3.3	+2.5
Other products	1.5	0.5	−1.0
Total industrial wood	8.7	10.7	+2.0

As part of its analysis of timber trends in the United States, the U.S. Forest Service [ref. 28] has attempted to project consumption levels for an equal period into the future. The projection was developed under the general assumptions that population would increase from 187 million in 1962 to 325 million in the year 2000, that gross national product at 1961 prices would increase from $546 billion in 1962 to $1,920 billion in the year 2000, and that prices of timber products would remain approximately stable relative to prices generally. The projected consumption levels, developed by the Forest Service in considerable detail, have been summarized briefly below.

Product	1962	2000 Projected	Change, 1962 to 2000
	Billion cubic feet, roundwood equivalent		
Sawlogs	5.9	8.4	2.5
Veneer logs	1.0	2.7	1.7
Pulpwood	3.3	9.2	5.9
Other products	0.5	0.5	—
Total industrial wood	10.7	20.8	10.1

With an allowance for fuelwood consumption, the Forest Service projects a rise in total wood consumption from 11.8 billion cubic feet in 1962 to 21.3 billion cubic feet in the year 2000. Import possibilities are reviewed briefly and a generally proportional expansion is considered possible. Possibilities for expanded exports are simply noted in passing. The projection indicates that net imports, which amounted to 14 per cent of total industrial wood consumption in 1962, will amount to 13 per cent of industrial wood consumption in the year 2000.

The resulting estimates of domestically supplied wood are then converted to estimates of cutting levels through allowances for changing utilization practices and amounts of logging residues. On this basis the total cut from the forests of the nation is projected as increasing from 10.1 billion cubic feet in 1962 to 21.6 billion cubic feet in the year 2000.

As was discussed earlier in this chapter, under such a trend in cutting levels and present trends in forestry, the volume of cut in cubic feet would equal the volume of growth by 1990. By the year 2000, the current 60 per cent surplus of growth over cut would have been changed to a 20 per cent growth deficit. Pronounced shortages in particular sizes, species, and regions would have been encountered earlier.

An evaluation of the Forest Service projections is beyond the scope of this study; some comments on these projections have been offered elsewhere [ref. 32]. The projections have been sum-

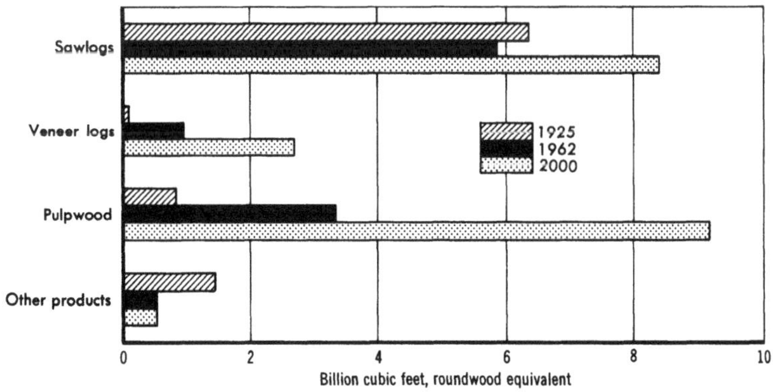

FIGURE 4. *U.S. wood consumption in industrial forest products, 1925 and 1962, and U.S. Forest Service projections for the year 2000.* (Source: *U.S. Forest Service data.*)

marized simply as part of the general background of the problem analysis. They represent a carefully developed indication of a possible line of development of the timber of the United States, but, as is stressed by the Forest Service, there is nothing inevitable about any such pattern of development. The forest industries will have to be much more effective in product development and marketing in the future than they have been in the past if they are to approach the domestic market levels indicated in the Forest Service study.

CHAPTER 3

THE U.S. IN THE WORLD BALANCE OF TRADE IN WOOD

In 1962, the United States had a total annual forest growth in cubic-foot terms some 60 per cent greater than the annual cut; yet in that same year net imports of forest products represented about 14 per cent of the wood equivalent of total forest products consumption (excluding fuelwood). In many ways this represents an unusual situation. The United States is the only one of the economically developed nations that combines a timber growth surplus with a deficit position in its wood balance of trade.

Today, the United States is the world's leading importer of forest products. The value of such imports in 1963 totaled $1.7 billion, or 23 per cent of the world total reported by the FAO [ref. 9]. The United Kingdom ranked second with imports of $1.3 billion, while the other Western European nations had total wood products imports of $2.8 billion. The three areas together accounted for 70 per cent of the total value of forest products imports.

In addition, the United States is an active exporter of forest products, ranking fourth among the nations of the world in value of such exports. In 1963, these exports totaled $0.7 billion, or 10 per cent of the world total reported by the FAO. The leading three nations are Canada ($1.7 billion), Sweden ($1.0 billion), and Finland ($0.8 billion), while the U.S.S.R. ranks fifth ($0.4 billion).

This chapter is devoted to a review of the major elements of the past and present participation of the United States in international trade in forest products. In keeping with the study's focus on the nation's timber resources, discussion here deals with the wood equivalent represented by these forest products rather than the dollar values. The cubic-foot volume of roundwood equivalent is used as the common unit of measurement for all of the forest products.

The Historical Development[1]

In the tradition of colonial areas, the early American colonies turned to the export of raw materials and semiprocessed goods to provide for imports of manufactured items from the British Isles and Western Europe. The abundant forests provided a ready basis for this export trade. As the young nation developed, it continued in this tradition as an exporter of forest products. However, the great demands of rapid internal expansion provided ready markets for the greater part of the output. The forest industries developed almost solely on the basis of domestic demand; exports became relatively less important. Thus, in 1899 lumber exports totaled 1.5 billion board feet while imports amounted to 0.7 billion board feet; net exports amounted to 2 per cent of the domestic consumption of 34.3 billion board feet. The youthful pulp and paper industry was already on a net import basis approaching 20 per cent of its wood equivalent.

At the beginning of the twentieth century, net lumber exports representing 175 million cubic feet of roundwood equivalent more than offset the 35 million cubic feet of net imports in the pulp and paper field. The resulting total net exports of 140 million cubic feet represented slightly less than 2 per cent of the total industrial wood production of the nation.

Net exports of lumber rose to an all-time peak of 320 million cubic feet in 1913, but then dropped sharply during World War I. Meanwhile, the net imports in the pulp and paper field continued to rise, and by 1915 became greater in roundwood equivalent than the net exports of lumber. The nation had become a net importer in forest products and has continued in this position to the present.

[1] Data reported in this section are as estimated by the U.S. Forest Service [ref. 27].

The twenties were marked by a strong recovery of the lumber export trade and net exports rose again. However, net imports in the pulp and paper industry rose even more rapidly, because the industry's primary dependence on spruce and fir had not yet been broken by the sulfate-pulping process. Thus, in 1929 net exports of lumber amounting to 255 million cubic feet were more than offset by net imports of 590 million cubic feet in the pulp and paper industry. Over all, the nation was a net importer by 4 per cent of its industrial wood use.

Lumber exports declined during the depression years of the thirties, while net imports in the pulp and paper field continued to mount. By 1939, net imports in the pulp and paper industry were 595 million cubic feet, compared to only 60 million cubic feet of net lumber exports. The nation had become a net importer to the extent of 8 per cent of its industrial wood use.

Under the impact of World War II, the United States lost its

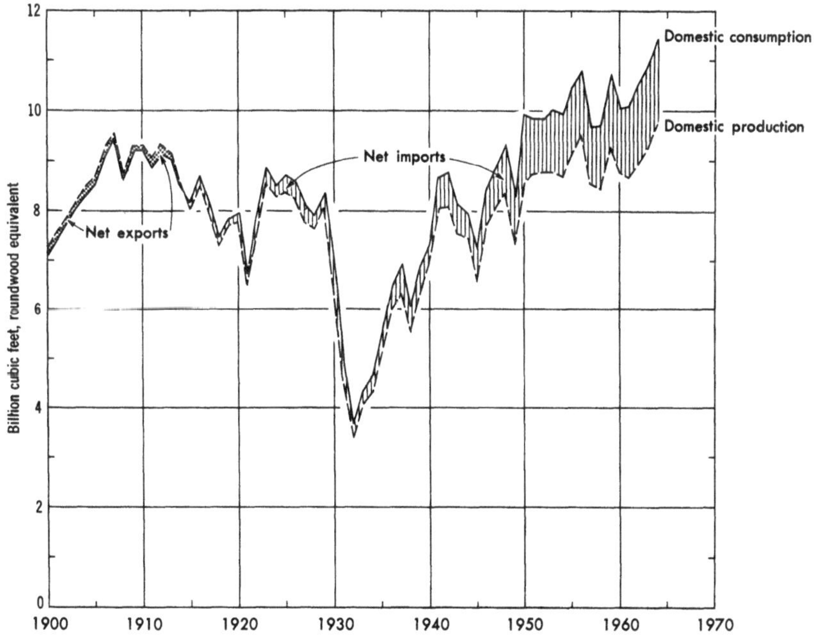

FIGURE 5. *U.S. production and consumption of industrial wood products, 1900–64.* (Source: *U.S. Forest Service.* The Demand and Price Situation for Forest Products, 1964. *U.S. Department of Agriculture Misc. Pub. No.* 983 [*Washington: U.S. Government Printing Office, 1964*].)

traditional role as a net exporter of lumber, and, with the exception of 1947, the nation has been a net importer since 1941. By 1951, net imports in the lumber industry represented 235 million cubic feet, net imports of the pulp and paper industry totaled 945 million cubic feet, and net imports of hardwood plywood and veneer were appearing. Total net imports of 1,190 million cubic feet represented 12 per cent of industrial wood use in 1951.

Net imports in the pulp and paper industry reached a peak in 1951 and then began to decline. However, steady increases in net imports in lumber and in hardwood plywood and veneer more than offset this decline. During the five-year period 1960–64, net imports averaged 1,486 million cubic feet annually, representing 14 per cent of industrial wood use during the period.

The 1963 Trade Balance in Wood

To enable a more detailed analysis of the current United States balance of trade in wood, data on import and export volumes for 1963 have been compiled and converted to volumes in roundwood equivalent (see Table 2). While some minor elements are not included in the table, all of the items traded in substantial volume are believed to be represented.

The imports run heavily to lumber and newsprint paper (41 per cent and 23 per cent respectively); various grades of wood pulp total another 20 per cent; pulpwood, 6 per cent; hardwood plywood and veneer, 7 per cent; and other items the remainder. In contrast, the exports are more dispersed, with softwood sawlogs representing 23 per cent of the total; softwood lumber, 19 per cent; various grades of wood pulp, 33 per cent; paper, paperboard, and products, 20 per cent; and other products the remainder.

Many of the commodity classifications are traded actively on both an import and an export basis. Of the commodities showing a net export surplus total net exports are 283 million cubic feet. Of this, softwood sawlogs represent 54 per cent; paperboard, 18 per cent; alpha and dissolving pulp, 15 per cent; paper products, 8 per cent; and paper grades other than newsprint, 5 per cent. The commodities showing net imports on balance have a total of 1,708 million cubic feet. Of this, softwood lumber represents 43 per cent; newsprint paper, 28 per cent; hardwood plywood and veneer, 8 per cent; bleached sulfate pulp, 7 per cent; pulp other than bleached sulfate and dissolving grades, 6 per cent; pulpwood, 6 per cent; and other items, 2 per cent.

TABLE 2. International Trade in Wood Products, United States, 1963

Item	Commercial unit	In commercial units			In million cubic feet of roundwood equivalent		
		Imports	Exports	Net imports[1]	Imports	Exports	Net imports[1]
Sawlogs, softwood	million bd. ft.	–	880	−880	–	150	−150
Sawlogs, hardwood	million bd. ft.	54	71	−17	9	12	−3
Lumber, softwood	million bd. ft.	5,032	743	4,289	855	126	729
Lumber, hardwood	million bd. ft.	303	134	169	49	22	27
Plywood, softwood	million sq. ft.	10	19	−9	1	1	–
Plywood, hardwood	million sq. ft.	1,621	2	1,619	115	+	115
Veneer, hardwood	million sq. ft.	1,468	79	1,389	22	1	21
Pulpwood	thousand cords	1,501	92	1,409	116	7	109
Wood pulp							
Alpha and dissolving	thousand tons	261	524	−263	42	85	−43
Bleached sulfate	thousand tons	1,240	500	740	181	73	108
Other	thousand tons	1,273	397	876	178	56	122
Paper							
Newsprint	thousand tons	5,413	118	5,295	476	10	466
Other	thousand tons	140	256	−116	20	36	−16
Paperboard	thousand tons	45	740	−695	3	52	−49
Construction paper and board	thousand tons	184	34	150	14	3	11
Paper products	thousand tons	38	191	−153	5	27	−22
Total					2,086	661	1,431

– Negligible.
[1] Balances representing net exports rather than net imports are indicated by a minus sign.

Source: Data in commercial units from U.S. Department of Commerce; conversions to roundwood equivalents made in course of study.

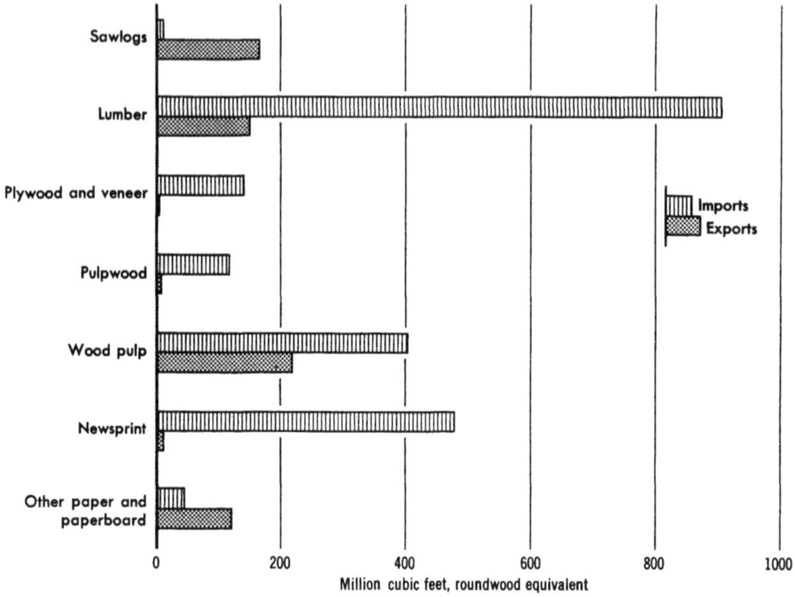

FIGURE 6. *International trade of the United States in wood products, 1963.* (Source: *Table* 2.)

The nature of recent developments in each of the more important commodity groups will be summarized below.

Softwood Sawlogs

Sawlogs are imported into the United States only in minor amounts, whereas sawlog exports represent nearly one-fourth of the roundwood equivalent of all forest product exports. The volume of such exports has risen rapidly in recent years and is currently dominated by exports to Japan. Although there is some uncertainty as to the exact quantities involved because of confusion in the various units used in reporting, Bureau of the Census statistics indicate that in 1957, U.S. exports of softwood logs totaled 107 million board feet, with 71 million board feet going to Canada and 34 million board feet going to Japan. By 1963, the total had risen to 880 million board feet, with 169 million board feet to Canada and 689 million board feet to Japan. While the exports to Canada had more than doubled, the exports to Japan had increased by 20 times.

The export of logs to Canada is a continuation of a well-established pattern. Much of these exports comes from timber adjacent to the border, often moving to Canadian mills, which are favorably located from the standpoint of transportation costs. Some use has been made of Small Business Administration set-aside procedures as a means of eliminating Canadian purchasers from public timber sales in this situation, but generally there has not been any pronounced reaction to this trade.

Far more attention has been given to the rapid rise in exports to Japan. More than 90 per cent of these exports has come from Washington and Oregon, with the balance being from California and Alaska. The U.S. Department of Commerce has estimated that 10 per cent of this log volume is in Douglas fir, 10 per cent in Port Orford cedar, and the remaining 80 per cent largely in hemlock and true firs. Thus, much of the export volume has been in species which tend to have weak domestic markets for sawlogs.

This rise in exports to Japan has been attributed to various factors, including rapid economic growth in Japan, the need for log markets following the 1962 blow-down in the Pacific Northwest, and even the Olympic Games. Correspondingly, these exports have been blamed for various ills of the forest industries of Oregon and Washington, which in reality seem to reflect a complex of forces. Some lumber manufacturers in the Pacific Northwest apparently believe that, if log exports were restricted, lumber exports to Japan would expand with the advantage of additional economic activity in the Northwest. The extent to which any enforced reduction of log exports to Japan would be picked up by expanded lumber imports by Japan is a debatable point, as is the question of whether any such increase in lumber imports would come from the United States or from other nations.

Some efforts have been made to foster political action designed to check this export trade. Efforts to prohibit the export of timber from lands owned by the State of Washington led to an analysis of the issue by the Battelle Institute [ref. 5] and to hearings before the State Land Commissioner. Formal application was also made for the blocking up of most of the National Forest lands of Washington and Oregon into a "sustained yield unit" under the provisions of the Cooperative Sustained Yield Act of 1944. The effect of this would have been to limit timber purchases to domestic producers. Hearings on this application were held in Washington and Oregon during 1965, but no such action was taken.

Softwood Lumber

In softwood lumber the patterns have been almost the opposite of those in softwood sawlogs. Export volumes have fluctuated fairly widely since World War II, but without any evident trend. From 1954 through 1963, exports tended to fluctuate between 550 and 650 million board feet, with a low of 550 million in 1958 and a high of 743 million in 1963. This high, however, is well below the figures reached in 1947 and 1951. In general, the level of softwood lumber exports has been not more than one-fourth of the level reached during the active export markets of the late twenties.

The levels of exports to individual countries have fluctuated sharply during the last ten years. Canada has usually been the largest single export market, but the level of these exports has declined markedly since 1959. In contrast, export volumes to Japan, Korea, West Germany, and Italy have moved upward. In 1963, Canada took about 14 per cent of these exports, compared with 15 per cent to Japan, 13 per cent to Korea, 12 per cent to Australia, 9 per cent to Italy, and 7 per cent to West Germany.

Softwood lumber imports have also fluctuated widely, but have shown a general upward trend, which has been particularly strong in recent years. During the late 1940's such imports varied between 1.0 and 1.5 billion board feet. By 1953, they had reached 2.5 billion board feet. Such imports rose to 3.3 billion board feet in 1955, dropped back to 2.7 billion board feet in 1957, and then rose steadily to 5.0 billion board feet in 1963. These 1963 softwood lumber imports amounted to 18 per cent of the domestic production of softwood lumber in that year.

Some 99 per cent of these softwood lumber imports come from Canada and correspond directly to lumber produced domestically. The rapid rise in these imports has come primarily from British Columbia, which supplied about two-thirds of the volume in 1962, with the balance coming largely from eastern Canada.

The reasons for this rapid rise have been attributed to many factors, including transportation cost advantages, the deflation of the Canadian dollar relative to the U.S. dollar, stumpage price policies, and labor cost differentials. The requirement of the Jones Act that U.S. flag vessels must be used for shipments between U.S. ports appears to place west coast mills at an appreciable cost disadvantage relative to coastal British Columbia mills in reaching east coast lumber markets. However, Canadian producers point out that by 1963 more than 60 per cent of the British Columbia exports originated in the interior forest zones. Differentials in stumpage prices have been difficult to evaluate because of differences in timber characteristics and associated logging costs. Rapidly rising Canadian wage rates appear to be removing any significant cost advantages in this area.

While a number of particular factors are undoubtedly involved in this development, the general movement is consistent with the entire historical pattern of development of the North American lumber industry. Since the early colonial days, the center of lumber production has tended to follow a retreating virgin forest frontier. The shifts of production from the Northeast to the Lake States, from the Lake States to the South, from the South to the Pacific Coast, and on the Pacific Coast from Washington southward into Oregon and California are all part of this familiar pattern. British Columbia now provides the most accessible remaining virgin timber opportunities and its expanding role in U.S. markets is a part of the whole pattern.

The rapid invasion of domestic softwood markets by British Columbia mills has, of course, been a matter of serious concern to a number of U.S. producers, particularly in Washington and Oregon. However, the domestic softwood lumber industry has by no means had a united front on the issue. A number of U.S. producers are also actively engaged in lumber production in British Columbia, and naturally have a different viewpoint than companies producing only in the United States. Further, some proposed actions would also shift the competitive position of Douglas-fir lumber producers relative to southern pine lumber producers in transportation costs and thus would involve one of the major splits within the domestic industry. An appeal for action under tariff legislation led to a study of the situation by the U.S. Tariff Commission [ref. 29], but the ruling was adverse to the appellant. Various other political actions have been suggested, but have received little general support.

Hardwood Lumber

Here, exports have shown no particular trend in the post-World War II period, fluctuating between 100 and 200 million board feet and totaling 134 million board feet in 1963. Canada is the primary market, taking 58 per cent of the total. As in the case of softwood lumber, the current export level is only about one-fourth that of the late twenties.

Hardwood lumber imports have shown some upward trend, rising from a level around 200 million board feet in the late forties to around 300 million board feet in recent years. In 1963, imports amounted to 303 million board feet. This was equal to 4 per cent of the volume of domestic production and was somewhat more than double the level of exports. Approximately, 43 per cent of this import volume came from Canada and corresponded to lumber produced in the United States. Japan, the Philippines, and Colombia were the leading sources of tropical hardwood lumber for the United States, but the three combined supplied less hardwood lumber to the United States than did Canada.

Softwood Plywood

Softwood plywood is produced in quantity only in the United States and Canada, and its use is largely restricted to these two nations. The 20 per cent tariff assessed against plywood imports by both the U.S. and Canada appears to have been effective in holding trade in softwood plywood between the two nations to very low levels. Producers in the United States have shown some interest in developing softwood plywood markets in Japan and Western Europe; in both instances this would require gaining acceptance for a material not now used in construction. In Japan, the lower grades of Philippine mahogany plywood fill some of the functions of softwood plywood. Similarly, in Western Europe the lower grades of beech and African hardwood plywoods and the rapidly expanding particle board production serve many of these functions. Thus, at present, softwood plywood is not an important item in international trade in forest products.

Hardwood Plywood

In 1963, U.S. market shipments of hardwood plywood amounted to 1.4 billion square feet surface measure, imports totaled 1.6 billion square feet, and exports were negligible. Thus imports amounted to 115 per cent of U.S. shipments and to 54 per cent of the total apparent consumption. This is the second highest ratio of imports to domestic consumption in any of the major forest products classifications, exceeded only by newsprint paper.

Six nations provided 94 per cent of these imports: 740 million square feet came from Japan, 273 million square feet from Taiwan, 247 million square feet from the Philippines, 120 million square feet from Korea, 78 million square feet from Finland, and 72 million square feet from Canada. The major species group represented is lauan, often termed "Philippine mahogany." In 1963, the imports included 1,168 million square feet of the so-called Philippine woods, 196 million square feet of birch, 134 million square feet of sen, and 123 million square feet in unspecified species.

Japan was the source of 40 per cent of the Philippine woods, all the sen, 25 per cent of the birch, and 58 per cent of the other species. Taiwan, the Philippines, and Korea produced primarily the Philippine woods. In contrast, practically all of the imports from Finland and Canada were birch, with these two nations and Japan being the primary sources of birch plywood imports.

This heavy import dependency is of relatively recent development. In 1951, imports amounted to 67 million square feet compared with 805 million square feet of domestic production, with the imports being primarily birch plywood from Canada. The first sharp rise in imports of the Philippine woods came in 1953, with total imports amounting to 21 per cent of U.S. consumption in that year. The Philippine mahogany plywood found a major market as door-skins for the flush-type panel doors now used extensively in U.S. construction. The rapid rise continued until the 1959 imports of 1.3 billion square feet represented 57 per cent of total U.S. consumption. Imports then dropped for two years before rising to the 1963 level of 1.6 billion square feet. Because of expansion in U.S. production, however, the relative position of these imports was less pronounced than in 1959.

To some extent these later years reflect the deliberate effort of Japanese producers to restrict the rate of expansion of exports because of fear of political action sought by U.S. hardwood plywood producers. Imposition of these export quotas, however, was accompanied by a rapid increase in imports from Taiwan, the Philippines, and Korea, and also by an increase in imports of hardwood veneer for lay-up in the United States. By 1963 veneer imports had reached 1.5 billion square feet by surface area, compared with only 0.2 billion square feet in 1949.

The Pulp and Paper Industry

The U.S. Forest Service [ref. 27] has estimated that the consumption of pulp and paper products in the United States in 1964 represented the equivalent of 3.5 billion cubic feet of roundwood volume. Of this total, about 22 per cent was provided by net imports, and the balance by domestic production. Of the total net imports, 17 per cent was in the form of pulpwood, while 83 per cent was in pulp, paper, or paperboard. Despite this heavy degree of over-all import dependency, the United States has an active and expanding export trade, especially in wood pulp and container board.

Pulpwood imports for 1964 are estimated at 160 million cubic feet against exports of 9 million cubic feet. This reflects a steady increase in net imports from a low of 100 million cubic feet in 1959, but is still well below the peaks reached in the past. Chip imports, which amounted to 20 per cent of the total in 1953, represented 54 per cent of the total in 1964. These chips are produced in British Columbia and exported to the Pacific Northwest. With the major expansion of the pulp industry now projected in British Columbia, a sharp reduction in the volume of these chip imports seems likely in the future. The balance of the material is roundwood from the provinces of eastern Canada, imported into the northeast and north central regions. The main exception is about 9 million cubic feet imported from the Bahamas into the southern United States.

Wood pulp imports totaled 2.8 million tons in 1963, equaling 9 per cent of the volume of domestic production. Exports amounted to 1.4 million tons, giving net imports of 1.4 million tons, or about 4 per cent of the total of production plus net imports. Although the total pulp imports have risen from 2.0 million tons in 1954 to 2.8 million tons in 1963, a more rapid rise in exports from 0.4 million tons to 1.4 million tons over the same period has resulted in a downward trend in the level of net imports.

The wood pulp exports in 1963 were concentrated in the special alpha and dissolving grades, and bleached sulfate pulp. Exports in the special alpha and dissolving grades amounted to 38 per cent of domestic production and were about double the level of imports. These exports go primarily and in about equal parts to Western Europe and the Far East. The level of exports to the Far East largely reflects the flow to Japan of production of a dissolving pulp mill in Alaska which is controlled by Japanese interests. In contrast, the exports of bleached sulfate were less than 9 per cent of production and went primarily to Western Europe, while imports were about two and a half times as great as exports. Similar surpluses of imports over exports occurred in the other grades. Of the total wood pulp exports in 1962, Western Europe was the market for 51 per cent, the Far East for 30 per cent, and Latin America for 11 per cent.

The wood pulp imports in 1963 were heavily concentrated in the bleached sulfate and bleached sulfite grades, in each case amounting to more than 20 per cent of U.S. production. Unbleached mechanical groundwood was next in volume, followed by the dissolving grades and unbleached sulfate. Canada was the dominant source of all the major grades imported, providing 87 per cent of the total in 1962. Western Europe provided most of the balance, mainly in the sulfate and sulfite grades.

Thus, the net import balance of the United States in wood pulp is essentially a Canadian relationship, with net imports from Canada in 1962 of 2.4 million tons. In contrast, the balance with the rest of the world involved exports of 1.1 million tons against imports of 0.3 million tons. In view of earlier relationships with Western Europe, it is notable that the pulp exports to that region of 0.6 million tons were about double the imports of 0.3 million tons.

In paper, board, and products, imports of 5.8 million tons were far greater than the exports of 1.3 million tons. The net imports of 4.5 million tons amounted to 10 per cent of production (39.0 million tons) plus net imports.

This balance was controlled by the heavy imports of newsprint paper, which totaled 5.4 million tons. Canada supplied 96 per cent of this, with the balance coming from Finland. The level of newsprint imports has been fairly stable since 1956, while domestic production has been increasing. However, in 1963, imports constituted 73 per cent of U.S. newsprint supply.

Building board, imported mainly from Western Europe, was a poor second to newsprint in imports, amounting to 0.2 million tons in 1963.

Excluding newsprint, the United States had an export surplus in the paper, board, and products field in 1963, with exports of 1.2 million tons against imports of 0.3 million tons. Container board provided half of the total exports.

Western Europe was the most important market area for U.S. exports of paper, board, and products, including newsprint, taking 42 per cent of the volume. Latin America ranked second with 20 per cent, Canada third with 14 per cent, and the Far East fourth with 9 per cent.

CHAPTER 4

CANADIAN TIMBER AND FOREST PRODUCTS

The dominant element in the entire pattern of trade, as shown in the preceding chapter, is the importance of Canada as a source of forest products imports for the United States. The 88 per cent of roundwood equivalent of all forest products imports provided by Canada in 1963 included the following percentages of major products groups:

Pulpwood	99%
Newsprint paper	96%
Other paper, board, and products	26%
Softwood lumber	99%
Hardwood lumber	43%
Hardwood plywood	4%
Wood pulp	88%

Canada was also an important customer for forest products exports from the United States, taking 14 per cent of the roundwood equivalent of all such exports in 1963. This included the following percentages:

Pulpwood	100%
All paper, board, and products	14%
Softwood sawlogs	19%
Softwood lumber	14%
Hardwood lumber	58%
Wood pulp	3%

In total, the wood equivalent of the imports from Canada is more than 20 times as great as the wood equivalent of the exports to Canada. In contrast, in its wood balance of trade with the rest of the world excluding Canada, the United States is a net exporter by a two-to-one ratio.

The approximately 1.8 billion cubic feet of roundwood equivalent supplied in 1963 constituted 88 per cent of such imports from all sources and were one-fifth as large as the 9.2 billion cubic feet of industrial roundwood produced within the United States in that year.

Important though Canada is as a source of wood supply for forest products consumption in the United States, the latter is even more important to Canada as a market for its forest products industries. In 1960, the total disposition of industrial wood in Canada was about 3.2 billion cubic feet, comprising 1.1 billion cubic feet consumed in Canada, 1.7 billion cubic feet exported in some form to the United States, and 0.4 billion cubic feet exported to other nations. Thus, exports to the United States accounted for 52 per cent of the total industrial wood handled in Canada in that year, and for some 52 per cent more volume than domestic markets in Canada.

These figures highlight the close interrelationships of the forest products industries of the two nations. The forest types of Canada are the northern extension of the forest types of the United States. The manufacturing technologies of the two nations are identical. Even a number of the larger forest products corporations are common to both sides of the border. While the border is a real one, with various trade barriers influencing the industrial structures and flows of trade of the two nations, Canada and the United States are to an appreciable degree, if not completely, a single economy in timber resource development.

Certainly the relationships are such that the timber resources of the United States cannot be adequately evaluated without direct consideration of Canada. The present chapter, therefore, is devoted to a brief review of Canadian timber and the Canadian forest products industries.

The Timber Base of Canada

Canada is a vast northern land with a total area of about 2.3 billion acres, which is approximately the same as the total area of the United States. In contrast, its 1962 population of 18.6 million people was only one-tenth that of the United States and was generally concentrated on the country's southern edge. Thus problems of access and inherent productivity are critical elements in any evaluation of land and land use in Canada.

According to estimates compiled by the Canadian Department of Forestry [ref. 2], the total productive forest land in 1963 amounted to 620 million acres: 331 million acres in the eastern provinces, and 289 million acres in the west. In more complete statistics for 1961, published by the Dominion Bureau of Statistics [ref. 3], the productive forest area was subclassified as consisting of 75 per cent of accessible area and 25 per cent of potentially accessible area. In the east, 71 per cent of the productive forest area was classified as accessible, while in the west the accessible area was estimated at 79 per cent of the total.

The significance that can be attached to this physical classification is uncertain. In an economic sense, accessibility of forest lands depends upon the value of the logs f.o.b. mill, and the costs of extracting and delivering the logs. Thus, accessibility varies with timber values and with the development of transportation systems and must be evaluated in more than a purely physical sense. A whole gradient of accessibility classes could be visualized, with each class being characterized by some level (either positive or negative) of economic rent at a given period. Nonetheless, even the arbitrary and simple classification used in these statistics raises the question of the extent to which these forests are, in reality, a part of the current resource base of Canada.

The total volume of merchantable timber in 1963 was estimated at 752 billion cubic feet, of which 36 per cent was in the east and 64 per cent in the west. Softwood volumes totaled 611 billion cubic feet, of which 68 per cent was in the west. Hardwoods totaled 141 billion cubic feet, with 65 per cent in the east. In the estimates for 1961, 82 per cent of the total timber volume was classified accessible and 18 per cent potentially accessible.

Approximately 64 per cent of the total volume was estimated to be in trees 10 inches and larger in diameter and thus classified suitable for sawtimber. The remaining 36 per cent of the volume was in trees from 4 inches through 9 inches in diameter. About 65 per cent of the softwood volume was in the sawtimber-size range, compared with 56 per cent of the hardwood volume.

Total utilization of these forests in 1962 has been estimated by the Department of Forestry at 3.4 billion cubic feet, including 3.2 billion cubic feet of industrial wood and 0.2 billion cubic feet of fuelwood. No comprehensive estimates of the volume of growth are available.

An estimate of the allowable cut from the Canadian forests, prepared by Love in 1961 [ref. 18], used 1959 inventory data to estimate the total allowable cut at 10.9 billion cubic feet. This included 7.1 billion cubic feet of softwoods on accessible forests, 2.3 billion cubic feet of hardwoods on accessible forests, 1.2 billion cubic feet of softwoods on potentially accessible forests, and 0.3 billion cubic feet of hardwoods on potentially accessible forests.

Evaluations of Canada's timber prospects have faced a problem similar to that in the United States: inadequate or incomplete forest survey data resulting in gross underestimates of the total inventory volume. Love noted that the 1959 inventory used in his analysis showed an increase of nearly 40 per cent over the 1956 inventory figures, with this increase resulting from improved estimates by a number of the provinces. The 1963 total volume estimate in turn is 10 per cent higher than the 1959 estimate. Presumably application of Love's method to these new inventory data would result in an estimate of allowable cut higher than 10.9 billion cubic feet. Thus Love's estimate that the actual cut could be increased 3.5 times without exceeding the allowable cut might reasonably be applied to the cutting level of 1962. Obviously, a wide variety of operating problems, including access, plant locations, and the sizes and species of materials utilized, would be encountered in approaching any such level.

A subsequent set of estimates for Canada published by the Food and Agriculture Organization [ref. 15] shows allowable cuts for the area operable at 1960 costs and prices of 7.5 billion cubic feet if managed for a mixture of sawlogs and pulpwood, of 8.3 billion cubic feet under similar assumptions except for a 10 per cent price increase, and of 11.1 billion cubic feet if all lands operable at 1960 costs and prices were managed on shorter, pulpwood rotations. Thus, the upper end of this range of estimates extends beyond Love's estimates.

Timber Utilization

The total production of industrial roundwood in Canada in 1960 was slightly more than 3.0 billion cubic feet. About 1.7

billion cubic feet of this was in logs and bolts, with 1.4 billion cubic feet being used in the manufacture of lumber, and 0.1 billion cubic feet going into veneer and plywood. Another 1.3 billion cubic feet was in pulpwood; this was supplemented by about 0.2 billion cubic feet of mill residues in providing the total pulpwood supply. The heavy concentration of the lumber and plywood industries in British Columbia is suggested by the fact that 62 per cent of the logs and bolts were produced in that province, with another 23 per cent coming from Ontario and Quebec. In contrast, 65 per cent of the roundwood cut for pulping came from Ontario and Quebec, 13 per cent from the Atlantic provinces, and 12 per cent from British Columbia.

Despite wide fluctuations in production, the Canadian lumber industry has shown some expansion, particularly in recent years. Total production declined from 7.9 billion board feet in 1955 to 7.1 billion in 1957, but then rose to 9.6 billion in 1963. Two-thirds of this 1963 production was sold in export markets. This recent expansion seems to have been keyed to U.S. markets, with the increase in production of 2.5 billion board feet from 1957 to 1963 being accompanied by an increase in exports to the United States of 2.4 billion board feet.

Within Canada, the expansion has been largely centered in British Columbia, where production increased from 4.4 billion board feet in 1957 to 6.4 billion in 1963. During the same period, British Columbia exports to the United States expanded from 1.8 billion board feet in 1957 to 3.6 billion board feet in 1963, with a greater increase shown in the interior than in the coastal region of the province.

The softwood plywood industry, which is centered in British Columbia, has shown a strong upward trend, with production rising from 0.4 billion square feet of ¼-inch equivalent in 1950, to 1.9 billion square feet in 1963. This industry is notable among Canada's forest products industries in that it is based primarily on domestic markets, with exports in 1963 taking only 15 per cent of total production. The hardwood plywood and veneer industry has also shown some expansion, but in 1960 the softwood and hardwood plywood industries combined took only 3 per cent of Canada's industrial wood utilization.

The Canadian pulp and paper industry takes about half of the total industrial wood utilization. This industry reflects a strong orientation to U.S. markets, the extensive forests of spruce and other desirable pulping species, and the availability of water power. Thus, the dominant elements in the industry are chemical wood pulp and newsprint paper produced for the export trade.

Total pulp production has expanded from 5.3 million tons in 1940 to 12.4 million tons in 1963; export markets took 27 per cent of this 1963 total. Chemical pulp production amounted to 6.2 million tons in 1963, with 49 per cent of this being exported as market pulp. Mechanical pulp production totaled 5.8 million tons, with only 4 per cent of this being exported as market pulp.

Newsprint production has increased from 4.7 million tons in 1940 to 6.6 million tons in 1963. Although production capacity in the industry continued to expand, there was little increase in actual production from 1956 through 1963, and in the later year the industry operated at 80 per cent of capacity. Approximately 92 per cent of the 1963 production was exported.

The production of all other paper and paperboard has increased from 0.8 million tons in 1950 to 1.25 million tons in 1963. This segment of the industry is oriented to domestic markets, with exports amounting to 31 per cent of production in 1963.

Geographically, the pulp and paper industry is centered in Quebec and Ontario, which together produced 7.7 million tons of pulp in 1962, compared with 2.4 million tons in British Columbia and 2.1 million tons in the other provinces. In paper and board, Quebec and Ontario produced 6.6 million tons compared with 1.2 million tons in British Columbia and 1.3 million tons in the other provinces.

Currently, projected expansions of the chemical pulp industry of Canada are the center of much attention from the worldwide chemical pulp industry. Proposals for some 21 new chemical pulp mills had been announced by the end of 1964, of which 17 were projected for British Columbia. According to an announcement of the British Columbia Bureau of Economics and Statistics [ref. 1], as of December 1, 1964, there were 14 existing pulp and paper mills in British Columbia, 3 new mills under construction, 5 proposed mills with timber supply secured, 3 proposed mills with applications for timber, and 6 proposed mills under investigation. If all 17 additions were to be realized, their capacity combined with announced increases in established mills would expand British Columbia chemical pulp capacity by about 3.5 million tons per year. The proposals for the rest of Canada represent nearly another million tons.

The Canadian Balance of Trade in Wood

An attempt has been made to approximate the Canadian wood balance of trade in terms generally similar to those used in the discussion of the U.S. balance of trade in wood. The estimates developed are presented in Table 3. Total forest product exports are estimated to be 2.3 billion cubic feet of roundwood equivalent while imports represent only 0.1 billion cubic feet. The net export balance represents two-thirds of the total volume of industrial wood handled by the forest products industries of Canada.

Approximately 80 per cent of the total exports in terms of wood equivalent is marketed in the United States. The most important secondary market is the United Kingdom, with which Canada enjoys British Commonwealth preference privileges. The total export volume includes 43 per cent in softwood lumber, 24 per cent in newsprint paper, 20 per cent in chemical pulp, 7 per cent in pulpwood, and 6 per cent in various other products.

Softwood lumber exports amounted to 4.5 billion board feet in 1955, dropped to 3.5 billion in 1957, and then rose to 6.3 billion in 1963. The United States took 77 per cent of this 1963 volume, the United Kingdom took 11 per cent, Japan took 5 per cent, Western Europe exclusive of the United Kingdom 2.5 per cent, and Australia 2 per cent. Canadian producers, recognizing the dangers in this heavy dependence on U.S. markets, emphasize that they are attempting to diversify their export markets. Possibly as a result of these efforts, exports to Western Europe have risen from 46 million board feet in 1960 to 160 million board feet in 1964, while exports to the United Kingdom have also moved upward during 1963 and 1964. Exports to Japan have risen markedly since 1960, reflecting the same developmental forces which have led to the rapid increase in softwood sawlog exports from the United States to Japan during this period. In spite of this increase, however, the United States remains the dominant market for the Canadian lumber industry. The apparent domestic consumption of both softwood and hardwood lumber in Canada in 1963 was only 3.6 billion board feet compared to total exports to the United States of 5.0 billion board feet.

Chemical pulp exports have risen with some fluctuations from 1.9 million tons in 1954 to 3.1 million tons in 1963. The U.S. share was 75 per cent in 1963. The United Kingdom received a little more than 7 per cent of the total in 1963, with the absolute

TABLE 3. International Trade in Wood Products, Canada, 1963

Item	Commercial unit	In commercial units			In million cubic feet of roundwood equivalent		
		Imports	Exports	Net exports[1]	Imports	Exports	Net exports[1]
Sawlogs, softwood	million bd. ft.	170	75	−95	29	13	−16
Sawlogs, hardwood	million bd. ft.	–	18	18	–	2	2
Lumber, softwood	million bd. ft.	130	6,328	6,198	17	987	970
Lumber, hardwood	million bd. ft.	110	155	45	16	24	8
Plywood and veneer	million sq. ft.	265	1,048	783	6	37	31
Pulpwood	thousand cords	140	[2]2,014	2,874	12	155	143
Wood pulp							
Chemical	thousand tons	74	3,042	2,968	12	456	444
Mechanical and other	thousand tons	–	298	298	–	24	24
Newsprint paper	thousand tons	–	6,212	6,212	–	547	547
Other paper and board	thousand tons	122	354	232	12	35	23
Total					104	2,280	2,176

– Negligible.
[1] Net import balances are represented by a minus sign.
[2] Including 900 thousand cords in chip exports to the U.S.

Source: Data on exports in commercial units from Canadian Department of Forestry; data on imports derived from Food and Agriculture Organization, *Yearbook of Forest Products Statistics, 1964* (Rome, Italy, 1964), and other sources; conversions to roundwood equivalents made as part of study.

volume being about the same as ten years earlier. In contrast, exports to European nations other than the United Kingdom doubled from 1960 to 1964, with such nations taking 8 per cent of the total in 1963. Exports to Japan have risen very rapidly since 1960, and in 1963 Japan took more than 5 per cent of the total exports.

Newsprint exports rose from 5.6 million tons in 1954 to a peak of 6.3 million tons in 1960, and then dropped back to 6.1 million tons in 1963. This was closely paralleled by the pattern of shipments to the United States, which took 4.9 million tons in 1954, 5.3 million tons in 1960, and 5.2 million tons in 1963. In 1964, newsprint exports rose to a new high of 6.8 million tons, with the United States taking 5.7 million tons. In 1963, the U.S. took 84 per cent of the total newsprint exports, the United Kingdom took 7 per cent, and the remainder was widely dispersed.

In addition to its dominant position as a market for Canadian lumber, market pulp, and newsprint, the United States also took 96 per cent of Canada's exports of hardwood plywood and veneer. The softwood plywood industry is the only exception

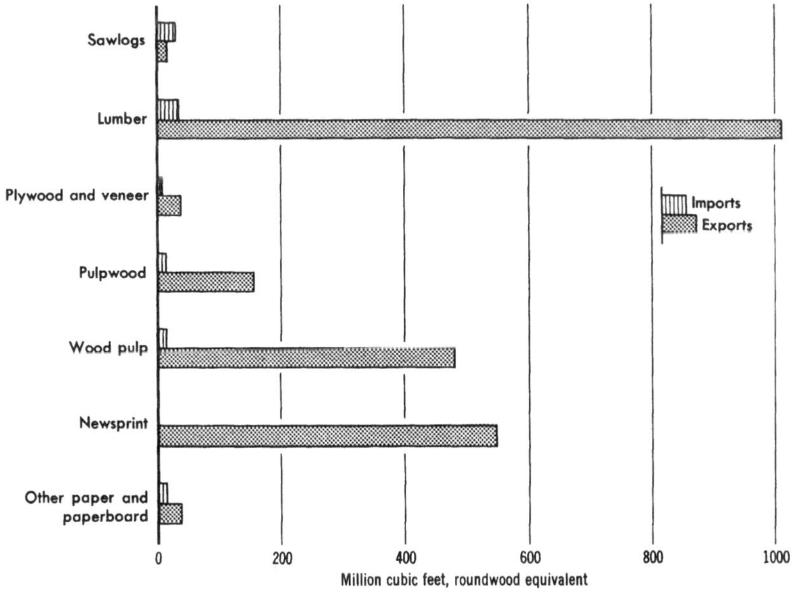

FIGURE 7. *International trade of Canada in wood products, 1963.* (Source: *Table 3.*)

to the pattern, with 91 per cent of the limited volume of exports going to the United Kingdom.

The United States is also the primary supplier of the relatively small volume of wood represented by Canadian imports, supplying Canada with practically all of its imports of softwood logs and lumber, with 90 per cent of the imports of hardwood lumber, with all of its small volume of pulpwood imports, with practically all of its wood pulp imports, and with most of its imports of paper, paperboard, and products. Only in its small consumption of tropical hardwood lumber, veneer, and plywood does Canada have any appreciable dependency on sources outside of North America.

CHAPTER 5

THE NORTH AMERICAN OUTLOOK

The previous chapters have presented reviews of the timber inventories, forest products outputs, and balances of trade in forest products of the United States and of Canada. In this chapter, the two nations will be considered together as constituting the North American region. This designation is in accordance with the practice of the FAO in its various analyses of forestry and forest products.[1]

With a combined population of 205 million in 1962, the two nations had 6.6 per cent of the world's population. However, according to estimates of the FAO [ref. 9], the North American region produced 35 per cent of the world's industrial roundwood removals in 1962, including 38 per cent of the softwood removals and 25 per cent of the hardwood removals. In this process, the region produced 29 per cent of the world's lumber, 59 per cent of the world's plywood, 56 per cent of the world's pulp, and 50 per cent of the world's paper and paperboard.

Obviously, North America is a major factor in the worldwide use of forest resources.

The North American Timber Base and Its Utilization

A few statistical highlights must serve to place the timber base of North America in broad perspective. Timber statistics

[1] A logical case can be made for also including Mexico as a part of this region. Geographically, it is part of North America and it is included in the North American Forestry Commission. In timber analyses, however, FAO includes Mexico with Latin America [ref. 7, 15, 16].

for the United States and Canada are not compiled on a fully comparable basis, but no appreciable distortions are believed to result from compilations in broad aggregates. Thus the North American timber base as of 1963 can be briefly summarized as follows:[2]

	United States	Canada	North America total
Forest area (*million acres*)	509	619	1,128
Volume of growing stock (*billion cu. ft.*)			
Softwoods	434	611	1,045
Hardwoods	194	141	335
Total	628	752	1,380
Volume in sawtimber-size range (*billion cu. ft.*)			
Softwoods	335	400	755
Hardwoods	101	79	180
Total	456	479	935

It may be of interest to compare this aggregate timber base with that of Europe, which is the other major consuming and producing region in forest products. The most recent estimates of the FAO [ref. 10] indicate that the total productive forest area of Europe[3] is 376 million acres, while the total volume of growing stock is 403 billion cubic feet. Thus, the productive forest area of North America is 3.0 times that of Europe, while the volume of growing stock in North America is 3.4 times that of Europe. In sharp contrast, the North American cut in 1962 was only 20 per cent larger than the 1959–61 average annual removal in Europe. (See Figure 8, p. 54.)

[2] The forest areas shown are the "commercial forest area" in the United States as defined by the Forest Service and the "productive forest area" in Canada as defined by the Dominion Bureau of Statistics. Growing stock volumes under U.S. definitions include volumes of trees 5 inches and larger in diameter; in Canadian statistics, such volumes include trees 4 inches and larger in diameter.

[3] Excluding the European part of the U.S.S.R.

The total cut in North America in 1962 was constituted roughly as follows:

	United States	Canada	North America total
	billion cubic feet		
Lumber production	5.1	1.7	6.8
Plywood and veneer production	0.8	0.1	0.9
Pulp and paper production	2.6	1.2	3.8
Other industrial use and export	0.5	0.2	0.7
Primary fuelwood use	1.2	0.2	1.4
Total cut	10.1	3.4	13.5

Some of the main elements of the forest products mix in 1963 were as follows:

	United States	Canada	North America total
Softwood lumber (*billion bd. ft.*)	27.6	7.4	35.0
Hardwood lumber (*billion bd. ft.*)	7.0	0.5	7.5
Softwood plywood (*billion sq. ft. 3/8″ equiv.*)	10.3	1.3	11.6
Wood pulp (*million tons*)	29.4	12.4	41.8
Newsprint paper (*million tons*)	2.1	6.6	8.7
Other paper and board (*million tons*)	36.9	1.3	38.2

In addition to the items listed, hardwood plywood and a wide variety of minor forest products were also produced.

The North American Balance of Trade in Wood

In considering North America as a single timber resource region, the question arises as to the volume of exports of forest products from the region to the rest of the world and the volume of imports from the rest of the world to the region. Estimates of these quantities are presented in Tables 4 and 5.

TABLE 4. Exports of Wood Products from the United States and Canada to Destinations Outside the North American Region, 1963

Item	Commercial unit	In commercial units			In million cubic feet of roundwood equivalent		
		U.S.	Canada	Total	U.S.	Canada	Total
Logs, softwood	million bd. ft.	711	42	753	121	7	128
Logs, hardwood	million bd. ft.	40	8	48	7	1	8
Lumber, softwood	million bd. ft.	633	1,701	2,334	99	265	364
Lumber, hardwood	million bd. ft.	55	26	81	8	4	12
Plywood, softwood	million sq. ft.	18	305	323	1	5	6
Plywood and veneer, hardwood	million sq. ft.	1	–	–	–	–	–
Pulpwood	thousand cords	–	510	510	–	39	39
Wood pulp	thousand tons	1,361	833	2,194	209	123	332
Alpha and dissolving	thousand tons	(512)	(86)	(598)	(82)	(14)	(96)
Sulfite	thousand tons	(206)	(217)	(423)	(31)	(33)	(64)
Sulfate	thousand tons	(638)	(458)	(1,096)	(96)	(69)	(165)
Other	thousand tons	(4)	(73)	(77)	–	(7)	(7)
Paper	thousand tons	320	1,060	1,380	38	98	136
Newsprint	thousand tons	(118)	(984)	(1,102)	(10)	(87)	(97)
Other	thousand tons	(202)	(76)	(278)	(28)	(11)	(39)
Paperboard	thousand tons	635	157	797	43	11	54
Paper products	thousand tons	150	4	154	21	1	22
Total					547	554	1,103

– Negligible.
( ) Indicates subtotal.

TABLE 5. Imports of Wood Products to the United States and Canada from Sources Outside the North American Region, 1963

Item	Commercial unit	In commercial units			in million cubic feet of roundwood equivalent		
		U.S.	Canada	Total	U.S.	Canada	Total
Logs, softwood	million bd. ft.	1	–	1	–	–	–
Logs, hardwood	million bd. ft.	46	–	46	8	–	8
Lumber, softwood	million bd. ft.	55	–	55	9	–	9
Lumber, hardwood	million bd. ft.	175	9	184	27	1	28
Plywood, softwood	million sq. ft.	5	–	5	–	–	–
Plywood and veneer, hardwood	million sq. ft.	2,350	260	2,610	120	5	125
Pulpwood	thousand cords	95	–	95	7	–	7
Wood pulp	thousand tons	243	12	255	36	2	38
Paper	thousand tons	230	–	230	23	–	23
Paperboard	thousand tons	183	–	183	13	–	13
Paper products	thousand tons	38	–	38	5	–	5
Total					248	8	256

– Negligible.

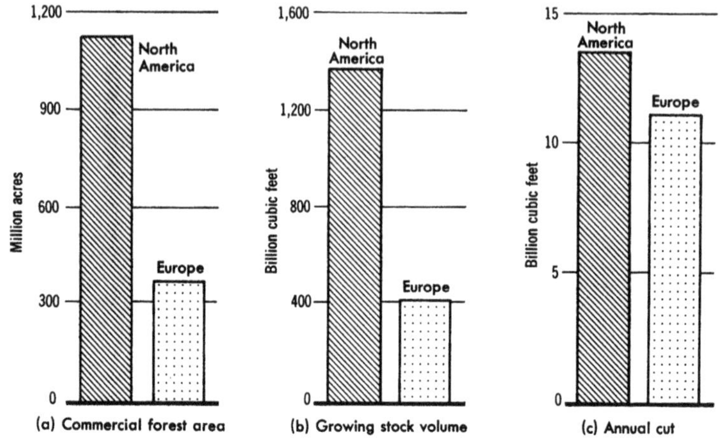

FIGURE 8. *Comparison of forest areas, timber volumes, and cutting levels in North America and Europe, 1962.*

Forest products exports from the region represented a roundwood equivalent of approximately 1.10 billion cubic feet in 1963. This was 8 per cent of the total cut for all purposes and 9 per cent of the cut of industrial wood in that year. In contrast, imports from outside the region totaled approximately 0.25 billion cubic feet of roundwood equivalent. Net exports, then, amounted to nearly 0.9 billion cubic feet, which represented 6 per cent of the total cut in the region and 7 per cent of the cut of industrial wood.

Softwood lumber represented 33 per cent of the total export volume, nearly three-fourths of which came from Canada; the balance came from the United States. Another 12 per cent of the total volume was represented by exports of softwood sawlogs, most of which came from the United States.

The pulp and paper industry provided a little more than half of the total export equivalent. Wood pulp shipments alone represented 30 per cent of the total export volume. Half of this volume was composed of sulfate pulp, with the remainder being primarily dissolving pulp and sulfite pulp. The United States was the source of three-fifths of the sulfate shipments, of more than four-fifths of the dissolving pulp shipments, and of about half of the sulfite shipments. Shipments of paper, board, and products amounted to 20 per cent of the total export wood equivalent. About two-fifths of this volume was in newsprint paper, coming largely from Canada. The United States was the primary source of the other categories in this general group, with container board being the most important single element.

The wood equivalent of the total export volume originated in approximately equal parts from each of the two nations. From Canada, the exports were heavily concentrated in softwood lumber, chemical wood pulp, and newsprint paper. Exports originating from the U.S. portion of the region were less concentrated, with chemical wood pulp, softwood sawlogs, softwood lumber and paper, board, and products other than newsprint all being of substantial importance.

Roughly 60 per cent of the limited volume of imports brought in from outside the region consisted of tropical hardwoods. These imports were brought in primarily in the form of hardwood plywood and veneer; second, in the form of lumber; and third, in a rather small volume of logs. The balance of the imports was largely in pulp and paper products. The U.S. portion of the region was the destination of most of these imports.

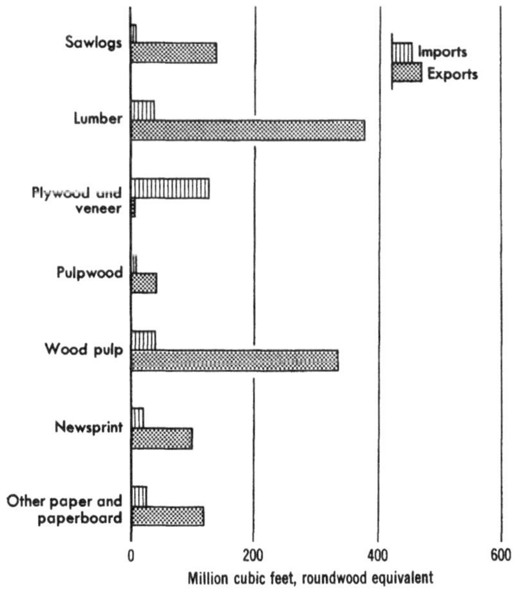

FIGURE 9. *North American trade in wood products with the rest of the world, 1963.* (Source: *Tables 4 and 5.*)

Consumption Trends and Forest Productivity

The apparent consumption of the United States in 1962 included 10.6 billion cubic feet of roundwood equivalent in industrial forest products and 1.1 billion cubic feet of primary fuelwood, giving a total consumption of 11.7 billion cubic feet. In Canada, apparent consumption of industrial wood products was about 1.2 billion cubic feet of roundwood equivalent, while primary fuelwood consumption amounted to about 0.2 billion cubic feet. In total, then, North American domestic consumption in 1962 included roughly 11.7 billion cubic feet of roundwood equivalent in industrial wood products and 1.3 billion cubic feet of primary fuelwood, giving a total of 13.0 billion cubic feet.

The total growth on commercial forest land in the United States in 1962 has been estimated at 16.3 billion cubic feet [ref. 28]. The allowable cut in Canada might be taken as more than 11 billion cubic feet. The total of annual growth in the United States, plus the allowable cut in Canada, is thus about 27.5 billion cubic feet—more than double the apparent domestic consumption of 13.0 billion cubic feet.

On the basis of detailed study, the U.S. Forest Service [ref. 28] has projected that the consumption of all industrial wood products in the United States in cubic feet of roundwood equivalent will, under stated assumptions, increase by 95 per cent from 1962 to the year 2000. This is in marked contrast to past experience over a period of equivalent length, in which industrial wood consumption increased by only 20 per cent from 1925 to 1962. The Forest Service also projected that primary fuelwood consumption would decline by 52 per cent from 1962 to the year 2000. No similar projections are available for Canada covering an equal period. If it is assumed for the moment that the percentage changes in Canadian consumption over this period can be taken as the same as those projected by the Forest Service for the United States, then the apparent consumption in Canada by the year 2000 would total 2.3 billion cubic feet of industrial wood and 0.1 billion cubic feet of fuelwood.

On this basis, total North American domestic consumption of wood products and primary fuelwood in the year 2000 might represent a roundwood equivalent of about 23.5 billion cubic feet. This would still be some 4 billion cubic feet below the estimated total of annual growth in the United States plus allowable cut in Canada in 1962. With a continuation of the present trends in forestry, this surplus over the projected total domestic consumption could readily rise to 6 billion cubic feet or more by 2000. The current net exports of forest products represent not more than 1 billion cubic feet of roundwood equivalent. Thus, even in the face of a very rapid rise in domestic consumption, the North American forests appear to have the physical potential to support a major increase in exports to the rest of the world over the remainder of this century.

As of 1962, imports to North America from the rest of the world represented only 0.2 billion cubic feet of roundwood equivalent. Of this, approximately half represents tropical hardwoods, while the remainder is in pulp and paper products which could readily be produced from North American timber. These tropical hardwood products represent appreciably less than 1 per cent of total domestic consumption of wood in North America.

Thus, currently available data appear to warrant the general conclusion that, aside from the less than 1 per cent of consumption based on tropical hardwoods, North American forests could sustain a 95 per cent increase in domestic consumption of industrial wood products and a 300–400 per cent increase in exports by the year 2000. To the extent that domestic markets may fail to develop as rapidly as suggested by the U.S. Forest Service projections, the opportunities and even the need for expansion of export markets would be even greater.

This general conclusion may seem to gloss over all questions of species composition, physical characteristics of logs, and economic location of the North American timber. Certainly there are major discrepancies between the elements of this total timber inventory and the corresponding elements of the timber actually cut and used in 1962. The very large surplus of current growth in small hardwoods, the decline in size and changing characteristics of the softwood sawlogs of the West, and the whole question of economic accessibility of an appreciable part of the Canadian timber are among the more obvious and important considerations here.

Most obviously, the forest products industries cannot continue blithely along the patterns of timber utilization of 1962 and achieve the kinds of expansion indicated from the timber base of North America. If, however, the forest industries are able to achieve continuing adjustments in their technologies and operations to the kinds of timber that are physically extant while holding their costs to a competitive level relative to alternate materials, then the timber base of North America can prove to be a resource capable of sustaining the magnitude of expansion indicated. Positive programs of forest management can contribute to this outcome.

Trade Barriers in North America

While the primary aim of this chapter is to consider North America as a single region in timber and forest products, it must be recognized that in fact the North American region consists of two independent nations, and that a variety of trade barriers serve to split the region to some degree. The major impediments to trade between the United States and Canada are the various import duties and import quotas. To these must be added various invisible trade barriers, of which the most important are the conventional escape clauses, overt nontariff protection such as arbitrary valuation of goods for duty purposes written into law, and unwarranted use of administrative authority to exclude imports (cf. [ref. 19]).

In general, the tariff schedules of the United States and Canada tend to favor the entry of raw materials or semimanufactured items and to be restrictive in varying degree to manufactured items. Such manufactured items as cabinets and furniture have complicated tariff structures.

Duties applicable to wood and paper and to printed matter are summarized in schedule 2 of the Annotated Tariff Schedules of the United States. Classifications are detailed and subject to varying rates. A few selected items, however, can serve to illustrate the patterns.

In the field of wood and wood products, logs enter duty-free. The preferential rate on softwood lumber is 25¢ per thousand board feet for eastern white pine and red pine; 35¢ per thousand for spruce; 75¢ per thousand for cedar; $1.00 per thousand for other pine, Douglas-fir, fir, hemlock, or larch; and $1.50 per thousand for other species. For most hardwood lumber originating in Canada, the rate is $1.50 per thousand. Hardwood flooring in strips or planks is subject to a preferential rate of 4 per cent ad valorem, while for other types of hardwood flooring (blocks, sectional units, etc.) the rate is 16⅔ per cent.

Plywood with a birch face enters at the preferential rate of 15 per cent ad valorem. For most plywood with a softwood face the rate is 20 per cent. For some faces (Spanish cedar, Paraná pine) rates go as high as 40 per cent.

In the pulp and paper field, pulpwood, pulp chips, and all grades of wood pulp enter duty-free. Newsprint paper is also admitted without duty. All other grades of paper and board are subject to various duties. For example, uncoated printing paper not subject to special classification enters at a rate of 17¢ per pound plus 4 per cent ad valorem, while most coated printing papers have a rate of 2¢ per pound plus 4.5 per cent ad valorem.

Canadian tariff schedules on forest products range from as high as to several times higher than U.S. tariff schedules, with rates going as high as 160 per cent on some classifications of paper.

The extent to which these rates are a significant barrier to trade appears to be highly variable. In the case of softwood lumber, the rates are relatively low and do not seem to be an important barrier to shipments from Canada to the United States. The rapid expansion of British Columbia softwood lumber exports to U.S. markets led to a petition for action under the escape clause of the Trade Expansion Act of 1962. This was rejected by the U.S. Tariff Commission [ref. 29] on the grounds that it had not been shown that the increases were attributable in major part to concessions made under trade agreements.

The 20 per cent duty on softwood plywood imports imposed by the United States is matched by a 20 per cent duty imposed by Canada. The possibility of some reduction of this level in the current GATT negotiations has been a source of apprehension to some U.S. plywood producers. The probable effect of any such reduction involves complex questions. Under the influence of the mutual tariff wall, the industries of the two nations have tended to have separate developments with little flow across the border. During recent years, prices have tended to be higher in Canada than in the United States. In fact, by 1964 the differential had reached a level at which there was some limited movement of U.S. softwood plywood into Canada. This suggests that the current situation involves more effective protection for the Canadian producers than for the U.S. producers. However, the possibility exists that a reduction in the duty would stimulate a structural change in the Canadian forest products industries, with a shifting of resources from lumber to plywood instead of the present limitation of production of plywood to levels consistent with administered pricing in the protected domestic market.

In the field of pulp and paper, the tariff structure appears to have been an important factor in determining industry patterns in the two nations. These tariff effects appear to complement and accentuate comparative advantages, and it would be difficult to separate these two considerations in explaining industry structure. In any case, the U.S. industry has developed largely in terms of self-sufficiency in all grades of paper and board except newsprint, while relying appreciably on imports of pulpwood, wood pulp, and especially newsprint paper. Correspondingly, the Canadian industry is heavily concentrated in market pulp and newsprint paper. At the present stage of development, it seems questionable that the U.S. duties are an important factor in the domestic market position of U.S. producers of paper and board. In contrast, Canada has a number of small, high-cost mills producing various grades of paper for domestic markets under the shelter of high duty rates. Thus a mutual reduction in duty rates would have the immediate effect of exposing these Canadian mills to the competition of large-volume U.S. producers. The longer term effect, however, could conceivably be some structural shift in the Canadian industry as the large-volume U.S. paper markets were made more available to Canadian producers.

The present study is not concerned with any detailed investigation of these questions. The issues are raised to illustrate the kinds of barriers which exist within the total timber economy of North America. These barriers appear to have some effect in limiting the degree to which the timber in each of the two nations is available to forest products consumers in the other. Nonetheless, the highly developed trade in forest products between the two nations suggests that the forces of interdependence are stronger than the divisive forces of trade barriers. In any case, these trade barriers are subject to reduction or elimination if developing resource stresses in forest products should become strong enough to make this politically desirable.

In total, the situation seems to be one which warrants the consideration of North America as a single region in long-term resource evaluations of timber and forest products.

CHAPTER 6

SOME ASPECTS OF WORLD DEMAND AND TRADE IN FOREST PRODUCTS

The previous section of this report has been directed to an analysis of the North American timber and forest products situation, considering the timber base, the forest products consumption patterns and trends, and the international trade patterns of the United States and Canada both separately and as a region. The conclusion has been reached that, although the forest industries of the region face major and difficult problems of adjusting to the kind and location of timber that will be available, the North American timber base has the potential to sustain a 95 per cent increase in industrial wood use in the United States and Canada from 1962 to the year 2000, while still permitting a several-fold expansion in total export volume to the rest of the world. The only significant form of import dependency for the region is in the consumption of forest products based on tropical hardwoods, which currently represents less than 1 per cent of the roundwood equivalent of total forest products consumption in North America.

From this conclusion, it follows that developments in the timber economy of the rest of the world are more likely to be important to evaluations of the North American timber resource as a source of demand rather than as an alternate source of supply for domestic consumption. The primary question appears to be the extent to which the rest of the world may provide market opportunities for, or market demands on, the North

American timber base. Any such demands on Canadian timber would affect evaluations of the U.S. timber resource almost as directly as would such demands on U.S. timber, since it would affect the availability of the total flow of products to consumers in the United States. A secondary question is the extent to which tropical hardwoods can be expected to continue to make a contribution to the total pattern of wood use in the United States.

This conclusion regarding the most probable nature of any world impacts on the North American timber resource will be reinforced by the discussion of world developments in this chapter. The world demand and supply prospects reviewed here demonstrate that there is very little likelihood of softwoods or temperate zone hardwoods from the rest of the world becoming of increasing importance in North American markets; instead, the outlook is for deficits in some of the major consuming regions of the world. The analysis will also show that tropical hardwoods are not likely to become of importance in North American markets, except in their current role as specialty products.

The present chapter will be directed primarily to some highlights of world trade in forest products and to a review of the outlook in some of the main consuming regions with emphasis on the demand aspect. Chapter 7 will place more stress on possible supply developments. Again it must be emphasized that the analysis here of the world timber economy is developed from a very special and restricted point of view: the possible relationships of worldwide developments to evaluations of the timber resources of the United States.

Highlights of the World Trade Pattern in Forest Products

According to the compilation of the FAO [ref. 9], the total reported value of forest products exports for 1963 is $7.3 billion, while the corresponding value of imports is reported as $8.2 billion. The discrepancy in import and export value totals is simply one of the constant discrepancies between quantities and values of international trade in forest products as reported by the exporters and by the importers, reflecting differences in classification, in bases for pricing, in completeness of records, and in similar factors.

The export total of $7.3 billion included $3.5 billion from European nations, $2.4 billion from Canada and the United States, $0.4 billion from the U.S.S.R., $0.1 billion from Latin

American nations, $0.2 billion from African nations, and $0.5 billion from Asian nations. In terms of commodity groups, 37 per cent was in paper, paperboard, and manufactures thereof; 34 per cent was in wood, lumber, and cork; 19 per cent was in pulp and waste paper; and 10 per cent was in wood and cork manufactures. The bulk of the exports go to nations within the same region as the exporting nation, with only limited amounts being involved in intercontinental shipments.

Canada was the largest single exporter of forest products, with the greater part of these exports moving to the United States. Sweden, Finland, and Norway together had 1963 exports of nearly $2.0 billion compared to the $1.7 billion from Canada; these Northern European exports largely moved southward in Europe. The United States ranked fourth in value of forest products exports with a 1963 total of $0.7 billion; these exports moved primarily to Europe and Asia, with an appreciable volume also going to Canada. The two North American and the three Northern European nations together accounted for 61 per cent of the total reported exports in 1963.

The U.S.S.R. has moved up to fifth position among the forest products exporters. This reflects in large part the rapid rise in lumber production in the nation, which by 1960 was producing 30 per cent of the world's softwood lumber output. The exports were wholly dominated by softwood lumber and moved primarily to Europe, with the United Kingdom alone taking almost one-third of the export volume.

The most important element in the Asian forest products exports is the trade which has developed around the export of "Philippine mahogany" plywood and veneer to the United States. Japan is the dominant producer, basing its production on logs exported from the Philippines and, to a lesser extent, North Borneo. South Korea, Taiwan, and the Philippines also export such plywood to the United States. In addition, Japan has become an exporter of paper and paperboard to other parts of the Asian region.

The African exports consist largely of tropical hardwood logs shipped from West Africa to Europe. Brazil is the main source of the limited volume of Latin American exports, with hardwood logs and lumber going to markets in Europe, the United States, and South Africa.

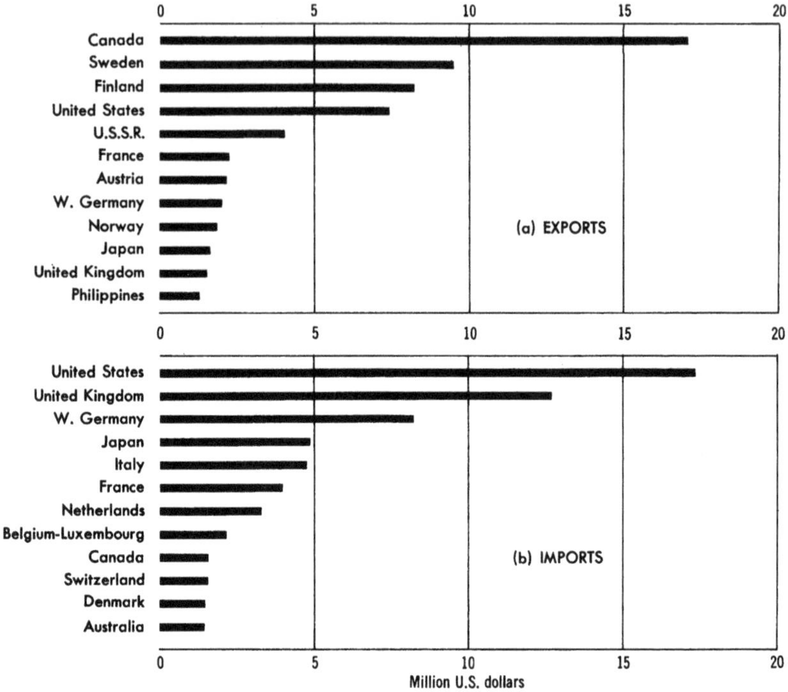

FIGURE 10. *Value of forest products exports and imports of nations leading in forest products trade, 1963.* (Source: *Food and Agriculture Organization.* Yearbook of Forest Products Statistics, 1964 [*Rome, 1964*].)

The reported import total of $8.2 billion included $4.5 billion by European nations, $1.9 billion by the U.S. and Canada, $0.9 billion by Asia, $0.4 billion by Latin America, $0.3 billion by Africa, $0.2 billion by the nations in the Pacific area, and less than $0.1 billion by the U.S.S.R.

The United States was the leading market with imports of $1.7 billion, while the United Kingdom was in second place with imports of $1.3 billion. The other Western European nations combined had imports of $2.8 billion. Thus the United States, the United Kingdom, and Western Europe accounted for 70 per cent of the total forest product imports.

In Asia, Japan received over half of the total value of imports, much of which was in the form of logs. Argentina was the largest single importer in Latin America, followed by Venezuela, Cuba,

Mexico, and Brazil. In the Pacific area Australia was the primary market.

It is of interest to note that six of the nations included among the twelve leading exporters of forest products—Canada, the United States, France, West Germany, Japan, and the United Kingdom—were also among the twelve leading importers. In five of these six nations the value of the imports was substantially higher than the value of the exports; the reverse was true only for Canada.

The European Outlook

The consumption of forest products in Europe (excluding the European portion of the U.S.S.R.) currently requires nearly one-fourth of the world's total volume of industrial wood removals. Although production is high and rising, the region is a net importer of wood on balance and provides the leading external market area for North American forest products exporters. Thus the European outlook is of direct relevance to evaluations of the U.S. timber resources.

A very thorough and detailed examination of timber trends in Europe since 1950 and projections of these trends to 1975 has been published by the FAO [ref. 10]; this study is the primary basis for the discussion presented here.

As is shown in Table 6, the consumption of forest products in Europe rose from a total wood raw material equivalent of 6.1 billion cubic feet in 1950 to 8.7 billion cubic feet in 1960 and is

TABLE 6. Total Consumption of Industrial Wood in Europe, 1950 and 1960, and Projections for 1975

(million cubic feet)

Product group	1950	1960	1975
Sawnwood	3,470	4,480	5,224
Wood pulp	1,165	2,280	5,295
Wood-based panel products	205	561	1,518
Other industrial roundwood	1,288	1,345	848
Total wood raw material	6,128	8,666	12,885
Use of residues	176	459	883
Total roundwood requirement	5,952	8,207	12,002

Source: Food and Agriculture Organization, *European Timber Trends and Prospects: A New Appraisal, 1950–1957* (Rome, Italy, 1964), Table 4, p. 82.

projected by the FAO as rising to 12.9 billion cubic feet in 1975. With allowance for the use of wood residues in these totals, the corresponding roundwood requirements are estimated at 6.0 billion cubic feet in 1950, 8.2 billion cubic feet in 1960, and 12.0 billion cubic feet in 1975.

Pronounced differences in trends are apparent in the data for the major product groups listed in Table 6. Sawnwood shows a steady decline in relative importance, while wood pulp is rising rapidly and is projected as requiring a volume equal to that of the sawnwood requirements in 1975. Wood-based panel products also show a strong upward trend, while other industrial wood products are moving down and are projected as declining even in absolute volume. In terms of the percentage of total wood material consumption, these changes of the major groups can be summarized as follows:

	1950	1960	1975*
	Per cent of total wood material		
Sawnwood	57	52	41
Wood pulp	19	26	41
Wood-based panels	3	6	12
Other wood products	21	16	6

*Projected.

These projections assume a 104 per cent increase in Europe's gross national product from 1960 to 1975 and stable relative prices for forest products. A separate projection under an assumed 87 per cent increase in gross national product yielded a total roundwood requirement of 11.0 billion cubic feet in 1975 rather than the 12.0 billion shown in Table 6. The FAO report stresses that neither figure is intended as a prediction of consumption, since it is recognized that actual price levels will depend upon the balance of supply and demand and thus may not be stable, as assumed in the projections.

In Europe, as in the United States, there has been a substantial replacement of lumber in its various end uses. Total consumption rose from 1950 to 1960 only because the expansion in economic activity was more rapid than this rate of displacement. The rise from 26 billion board feet to 32 billion board feet over this period represented an increase of 22 per cent, with a further increase of 17 per cent projected from 1960 to 1975. In con-

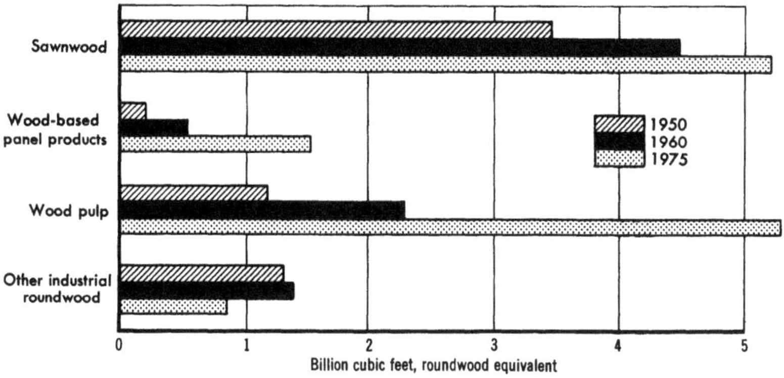

FIGURE 11. *European wood consumption in industrial forest products, 1950 and 1960, and FAO projections for 1975.* (Source: *Food and Agriculture Organization.* European Timber Trends and Prospects, A New Appraisal, 1950–1975 [*Rome, 1964*].)

trast, the consumption of paper and paperboard more than doubled from 11.4 million tons in 1950 to 23.3 million tons in 1960, and a further rapid expansion to 55.6 million tons is projected by 1975. The consumption of the various wood-based panel products has shown an even more rapid relative expansion.

As would be expected from the similarity of the broad economic forces controlling these trends in Europe with those in the United States, the nature of the shift in the forest products mix in Europe from 1950 to 1960 has generally paralleled that in the United States, and the further shifts projected by the FAO for Europe correspond broadly to those projected for the United States by the U.S. Forest Service [ref. 28]. In both areas, lumber has been declining in relative importance; paper and paperboard consumption has been rising rapidly; consumption of plywood and other wood-panel products has been rising even more rapidly; and other industrial wood products have declined in relative importance.

Data for the United States and Europe are not fully comparable in either classifications or time periods, but some rough comparisons can be made as follows:

1. In Europe, lumber consumption increased 22 per cent from 1950 to 1960; an increase of 17 per cent is projected from 1960 to 1975. In the United States, sawlog consumption decreased by 8 per cent from 1952 to 1962, but an increase of 16 per cent is projected from 1962 to 1980.

2. In Europe, wood pulp consumption increased 96 per cent from 1950 to 1960; an increase of 132 per cent is projected from 1962 to 1975. In the United States, expanding paper and paperboard consumption represented an increase in pulpwood equivalent of 23 per cent from 1952 to 1962; an increase of 63 per cent is projected from 1962 to 1980.

3. In Europe, the consumption of wood-based panel products increased by 174 per cent from 1950 to 1960; an increase of 170 per cent is projected from 1960 to 1975. In the United States, increases in plywood and veneer consumption represented an increase in equivalent plywood log volume of 114 per cent from 1952 to 1962; an increase of 190 per cent is projected from 1962 to 1980.

4. In Europe, the consumption of other industrial roundwood increased only 4 per cent from 1950 to 1960; a decline of 37 per cent is projected from 1960 to 1970. In the United States, the consumption of miscellaneous industrial wood declined 33 per cent from 1952 to 1962, but stable consumption levels are projected from 1962 to 1980.

This changing nature of the forest products mix in Europe is naturally accompanied by a shift in the kinds of wood required for processing. The primary element in this is a progressive and rapid increase in the role of small-sized roundwood relative to large-sized material. The FAO has estimated this shift to be about as follows:

	1950	1960	1975
	million cubic feet		
Large-sized roundwood	3,880	4,940	6,000
Small-sized roundwood	2,070	3,260	6,000
Total roundwood requirement	5,950	8,200	12,000

It is particularly interesting to note that 2.7 billion cubic feet of the projected increase of 3.8 billion cubic feet can be filled by small-sized roundwood, while only 1.1 billion cubic feet will require large-sized roundwood. Such a shift in the proportion of wood sizes is of obvious importance both in forest management and in harvesting operations.

From 1949–51 to 1959–61, Europe's recorded roundwood production for all purposes increased from 10.3 billion cubic feet to

11.2 billion cubic feet, while industrial wood production alone increased from 6.1 billion cubic feet to 7.5 billion cubic feet. The greater increase in industrial than in total wood production reflects a decline in the output of fuelwood. This increase of 1.4 billion cubic feet of industrial wood production was exceeded, however, by the increase of nearly 2.3 billion cubic feet in the roundwood equivalent of forest products consumption. By 1960, Europe had become a net importer of forest products to the extent of about 7 per cent of its industrial wood requirements.

To some extent, this represents a return to the situation which existed prior to World War II. During 1935–38, Europe was a net importer of about 0.35 billion cubic feet of roundwood equivalent annually. This import dependency was based largely on softwood lumber and logs (including pulpwood), for which the U.S.S.R. was the leading source. In pulp and paper, Europe was an appreciable net exporter.

In the immediate postwar period, trade was limited by controls in the United Kingdom and reconstruction needs in West Germany and the U.S.S.R. Thus, in 1949–51, Europe was actually a net exporter of wood products by 0.18 billion cubic feet of roundwood equivalent.

The following decade was marked by a rapid rise in imports, primarily in lumber from the U.S.S.R., pulp and paper from North America, and tropical hardwood logs from Africa. Exports increased to a lesser degree, and by 1959–61 Europe was a net importer to the extent of nearly 0.6 billion cubic feet. Unlike the prewar situation, however, Europe's new net import position was derived not only from logs and lumber, but also from pulp, while the export surplus in paper was rapidly disappearing.

In an effort to evaluate the ability of Europe to meet the projected rising forest products consumption levels from domestic sources, the FAO made detailed studies of all European countries and then aggregated the findings by six regions. The results indicated that both in 1950 and 1960, the northern region had a large surplus, the British Isles and the European Economic Community an even larger deficit, Central Europe a small surplus, and Southern Europe and Eastern Europe small deficits (and relatively low consumption levels). The analysis also indicated that all regions have the possibility of expanding output by 1975, with a total increase from 7.5 billion cubic feet in 1959–61, to 9.5 billion cubic feet projected in 1975. Nearly 1.3 billion cubic feet of this estimated increase would be in hardwoods, 0.7 billion cubic feet in softwoods.

The projected rise involves emphasis on thinnings and various marginal operations in which labor requirements per unit of volume tend to be high. In view of the rapid rise in labor costs and the growing difficulties experienced in recruiting woods labor, achievement of the projected increases without a pronounced advance in delivered wood costs will be difficult.

Bringing together its estimates of future consumption levels and possible cutting levels, the FAO study [ref. 10] states as follows:

> The expected output of industrial roundwood in 1975 of 270 million cu. m. [9.5 billion cubic feet] would fall well short of even the lower level of requirements, and at the higher level the shortfall would amount to the formidable margin of 70 million cu. m. [2.5 billion cubic feet] of roundwood a year by 1975. . . . What are the implications of a prospective shortfall in supply of this magnitude? There will not be any gap between supply and demand in 1975; sufficient will be forthcoming, either from additional output in Europe or from higher imports from other parts of the world, to meet requirements. But if this increased supply can be called forth only by a sharper rise in prices than has been assumed in the earlier estimates of requirements, this would eliminate some marginal demand for most forest products, so that requirements would be lower.

Obviously, primary interest should be attached to the general pattern shown rather than to the exact magnitudes. Europe has shifted from a modest net export position of about 0.2 billion cubic feet of roundwood equivalent in 1949–51 to a net import position of nearly 0.6 billion cubic feet in 1959–61. By 1975, a continuation of present patterns may lead to net imports estimated by the FAO as ranging from 1.5 to 2.5 billion cubic feet. While current net imports are dominated by lumber, only limited expansion of lumber imports is foreseen. The major import expansion indicated is in pulp and wood-based panel products, both of which can be manufactured from small-sized wood. With Europe expected to continue to expand its use of domestic hardwood species, the particular need or opportunity indicated is in the long-fibered pulps made from softwoods.

In terms of trade opportunities and demands on North American timber, the FAO analysis can be summarized by the following excerpts:

> The gap between Europe's demand for and domestic supply of sawn softwood is unlikely to widen materially by 1975, if at all. . . . The USSR and Canada are likely to remain the dominant sources of supply. There is no indication yet that sawn softwood production in the USSR, which doubled between 1950 and 1960, is reaching a limit. . . . Given the demand, Soviet exports to Europe could increase by some 50% to more than 1.5 million standards [3.0 billion board feet]. Canada could also increase its sawn softwood exports to Europe considerably. . . . Wood pulp and its products will constitute the largest item in Europe's increasing requirements of wood products, but Europe's supply of broadleaved pulpwood, and of the short-fibre pulp made from it, is likely to be adequate. . . . The shortfall in supply will be of coniferous wood, or of the long-fibre pulps that are made from conifers. . . . The deficit in Europe's long-fibre needs by 1975 is likely to be about 35 million cu. m. [1.2 billion cubic feet] of roundwood requirements.

It should be noted that the growing deficit of pulp and pulp products projected differs from the short-term movement of the last three or four years. A surge of new capacity has led to a condition of oversupply in some areas and an export surplus in chemical pulps; as this expanded output is absorbed by rising domestic markets by the late 1960's, the projected deficit is expected to develop rapidly.

From the nature of these projected deficits, it follows that North America and the U.S.S.R. will be the major possible sources for additional supplies. Although recognizing that the U.S.S.R. faces a great expansion in its own consumption of wood pulp products, the FAO indicates that planned expansion within the U.S.S.R. should permit substantial exports in 1975. Nonetheless, it seems likely that North America will be the major source. The question of supply sources other than North America will be developed more fully in the next chapter.

In view of the existing structure of the pulp and paper industry of Europe and the general availability of some form of pulping material, the FAO study suggests that imports are likely to be primarily in long-fiber market pulps and in paper grades with a high content of new wood pulp such as newsprint and kraft liner. The nature of the import pattern and its balance between pulpwood, wood pulp, and paper and board can, of course, be significantly influenced by tariff policies and other trade restrictions. Within Northern Europe there are strong economic pressures on the pulp producers to integrate vertically into paper and board as a means of expanding the value of output without corresponding increases in the volume of timber cut. The final location of expanded paper and board capacity in Europe is, however, subject to a variety of considerations, with the relationship between the "outer seven" and the "inner six" being a central source of uncertainty. These same uncertainties as to political elements in the trade patterns apply in even greater measure to potential expansion in imports from outside the total European region.

It must be noted that the FAO study, which is the primary basis of the discussion here, defines the level of import needs or opportunities from outside the region through a rather mechanistic approach to estimating the gap between European consumption and production levels. The assumption made is that the European requirement will be filled domestically to the full extent that silvicultural cutting levels will permit. The trade opportunities are estimated simply as the difference between the projected consumption levels and these estimated domestic production maxima. These limitations of the study model are clearly recognized in the published report.

From the standpoint of North America and other potential sources of supply outside Europe, the question of comparative costs deserves consideration. If these extra-European sources can supply a part of expanding European consumption at lower cost than European sources, then the volume of such imports to Europe might well expand even though European production has not yet closely approached the silvicultural maximum. Indeed, the current expansion in imports from non-European sources indicates that exactly such a movement *is* taking place.

It is at least conceivable (and perhaps even likely) that, in view of rapid increases in labor costs in Europe and the barriers to mechanization in some of the opportunities for expanded cutting levels, North America could supply not only the expanding long-fiber pulp needs at lower cost than that of the suggested marginal expansions of European output, but also an appreciable part of the hardwood pulps. The magnitude and accessibility of the hardwood timber inventory of the United States are such that a study of comparative costs between the U.S. and Europe in hardwoods would be of interest.[1] Again, it must be recognized

[1] The often-raised question of the potential role of tropical hardwoods in such competition is discussed in Chapter 7.

that possibilities of this kind depend not only on comparative costs but also on governmental trade policies.

This FAO study of timber trends is the only complete analysis of European timber and forest products consumption available. No thorough critical review of the findings has been published. Some questions have been raised concerning the projections of upward trends in lumber consumption, which appear to have a weaker statistical basis than the projections in the pulp and paper sector. During the first two years following completion of the study, lumber consumption actually rose more rapidly than projected. This reflected levels of construction well above those used in developing the projections. The question as to whether these higher current construction levels will result in lower needs for future construction and lower lumber consumption is an example of the kinds of uncertainties which surround such studies.

The projections in pulp and paper appear to be generally accepted. A new detailed study of European trends in the supply of and demand for pulp and paper has been undertaken in the Institute of International Economic Relationships, The University of Stockholm. Preliminary indications are that this study will support a somewhat higher projection in the consumption of pulp and paper in Europe than that presented in the FAO report. The study also involves an attempt to estimate future production levels from European sources, thus again giving a measure of market opportunities for extra-European sources as indicated by the gap between European consumption and European production. Publication of this study should provide further evidence of possible long-term impacts of European markets on North American timber.

As these concluding comments indicate and as the authors of the FAO report fully recognize, actual developments may vary widely from the projections which have been summarized here. Nonetheless, in its broad elements this report is a useful indicator of the kinds of interactions possible between European consumers and the North American timber resource.

The Western Pacific

In terms of current trade in forest products, Japan occupies a central position, not only in the Western Pacific region but in all of Asia (excluding the U.S.S.R.). The data compiled by the FAO [ref. 9] indicate that forest products exports from Japan in 1963 represented 34 per cent of the value of such exports from all of Asia, while forest products imports by Japan represented 51 per cent of the total value of all such imports by Asian nations. From the standpoint of the present study, this trade is of special interest because of its strong orientation to North America.

Japan provides a highly interesting example of rapid economic growth for a population which is very large relative to the nation's land base. With a population of 95 million people and a land area of 91 million acres, Japan has five times the population of California on a land area only 90 per cent that of California. Furthermore, the islands are entirely mountainous with no large plains.

Much of the area is forest land, with the total forest land area being estimated at 57 million acres. Only a little more than half of this forest area is classed as being in use. Although cutting on the area in use is heavy, domestic production is less than use and the nation is a net importer of forest products.

A report of FAO [ref. 6] indicates that the production of industrial wood in Japan in 1955 amounted to 1.4 billion cubic feet of wood. While the cut from the forests now in use is expected to drop sharply, Japanese authorities are reported to believe that the total production of industrial wood can be raised to 1.9 billion cubic feet by 1975 through opening up of the areas not now in use, development of plantation forestry, and similar measures. Even with allowances for pronounced substitution of other materials for forest products because of supply problems, a 1975 consumption level of 2.15 billion cubic feet is indicated. Thus a 1975 deficit of at least 0.25 billion cubic feet is projected as an indication of possible future net import levels.

This projection of expanding import dependency in industrial wood for Japan is supported by the experience of recent years. Japanese statistics as summarized in a U.S. Department of Commerce publication [ref. 23] indicate that total log consumption by the forest industries in Japan in 1963 amounted to 14.0 billion board feet, of which 3.0 billion board feet were imported logs. Thus, one-fifth of this total log use was apparently based on imports. These imports included 1.7 billion board feet of lauan, with four-fifths coming from the Philippines, and the balance from North Borneo. North America was the source of 0.7 billion board feet of logs, practically all softwoods. The U.S.S.R. supplied 0.4 billion board feet of logs, largely softwoods but with some birch included.

It is interesting to note that, although the volume of logs from North America was 65 per cent greater than the volume from the U.S.S.R., the value in North American log imports was three times that of the logs from the U.S.S.R. The value of Japan's imports of softwood logs from the United States rose from $7 million in 1958 to $88 million in 1963, while such imports from Canada rose from $0.2 million to $10 million over the same period. In contrast, softwood log imports from the U.S.S.R. rose from $7 million in 1958 to $34 million in 1963.

Japan's imports of softwood logs and lumber combined have risen spectacularly from 0.30 billion board feet in 1960 to 1.43 billion board feet in 1963. The North American share of this rose from 0.14 billion board feet in 1960 to 1.01 billion board feet in 1963. Softwood log imports from the United States rose from 0.80 billion board feet in 1960 to 0.52 billion board feet in 1963, while imports from Canada rose from a very low level to 0.1 billion board feet over the same period. Softwood lumber imports from the United States rose from 0.06 billion board feet in 1960 to 0.11 billion board feet in 1963, while imports from Canada rose from an almost negligible level to 0.26 billion board feet over this period.

The 2.6 to 1 ratio of logs to lumber in these 1963 imports apparently reflects greater profit opportunities from processing logs in Japan rather than importing lumber. The Japanese government permits the import of both softwood logs and lumber without duty and without any quantitative import controls.

Japan is also an importer of pulp and paper from North America. In 1962, the United States exported 0.35 million tons of wood pulp and 0.09 million tons of paper and board to the Far East, with Japan being the dominant market in this region. Canadian exports of pulp to Japan increased from 0.02 million tons in 1960 to 0.18 million tons in 1964. In addition, Canada exported 0.07 million tons of newsprint and 0.01 million tons of other paper and board to Japan in 1964. Japan also imports appreciable quantities of wood chips for pulping from the West Coast of North America.

With the continued economic growth of Japan, further increases in the total level of imports of these various products based on softwood timber seem probable. The primary alterna-

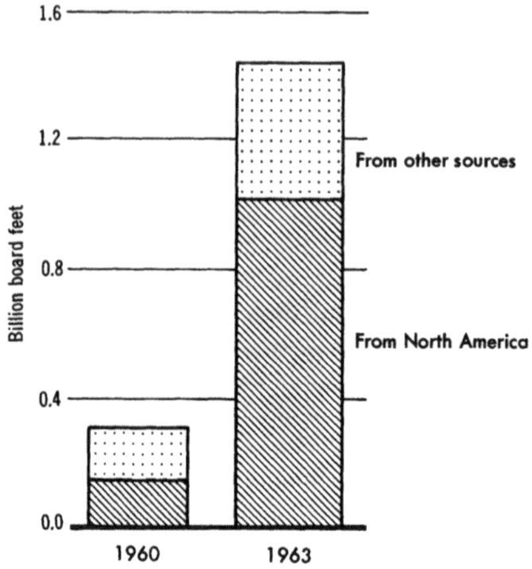

FIGURE 12. *Japanese imports of softwood logs and lumber, 1960 and 1963.* (Source: *U.S. Department of Commerce, Far Eastern Division, Bureau of International Commerce. "The Japanese Market for Softwood Logs and Lumber." 1964* [*mimeo*].)

tive to North America as a source of such imports is the Soviet Union. In 1963, Soviet exports to Japan, totaling 0.40 billion board feet, were primarily in the form of logs. Approximately one-eighth of this volume was used for pulp, with the balance being used in lumber manufacturing.

Economic development of the extensive forests of Eastern Siberia by the U.S.S.R. must almost surely be based in large part on export markets in the Asian region. The domestic population of the region is too low to make full use of the forest products potential of the region, while transportation problems are a major barrier to exports to the European market region. The two main export market possibilities seem to be Communist China and the nations of the Western Pacific. Currently the limited evidence suggests that the main emphasis is likely to be in the Western Pacific region. Both the timing of any such development and its impact on export market opportunities for producers in western North America are highly uncertain.

Approximately 0.82 billion board feet of lauan logs imported by Japan were used in 1963 in the manufacture of plywood and veneer. These imported logs are the primary basis of the Japanese plywood and veneer industry, although there is an appreciable volume of production based on domestic logs, primarily sen and birch. According to the Forestry Agency of Japan [ref. 17], slightly more than 16 per cent of the total volume of plywood and veneer production in 1963 was exported. Data obtained from the Japan Plywood Exporters Association show that during the first eleven months of 1964 shipments to the United States represented 71 per cent of the total export volume, with Canada taking another 7 per cent.

In recent years, the production of hardwood plywood and veneer from lauan logs has also expanded in the Philippines, Taiwan, and Korea. However, Japan remains as the primary producing area.

In large part, the Japanese plywood industry is based on importing logs, manufacturing them into veneer and plywood, selling the upper grades in export markets (primarily the United States), and disposing of the much greater volume of lower grades in domestic markets. Since the greater part of its raw material is imported from the Philippines, the industry is highly exposed to any decline in log exports from there, either as a result of governmental export controls or as a result of competition from domestic producers in the Philippines. Recognition of this danger and some problems of declining log quality appear to explain the recent interest in developing log sources from North Borneo and an even more recent interest in New Guinea.

Exports of forest products by the Philippines remain heavily oriented to logs. From 1961 to 1964, the value of log exports rose from $85 million to $135 million, while plywood and veneer exports rose from $11 million to $28 million, and lumber exports rose from $7 million to $8 million. The rate of expansion of the domestic plywood industry appears to be controlled both by the availability of capital and by the problem of finding markets for the large proportion of lower grade material as well as for the upper grades exported to North America.

If and when Philippine production does rise to a level such that log exports to Japan and other nations begin to decline, the effect could be disruptive to the Japanese plywood industry. From the standpoint of the United States as an importing nation, however, it would appear to make little difference whether the logs are manufactured in either the Philippines or Japan.

No reliable information could be obtained on the extent to which the lauan forest reserves are sufficient to maintain present production levels or to permit further expansions in the future.

Since the industry is now based on old-growth logs, it seems probable that increasing constraints on supply will be felt at some future period.

Mainland China

There is less known in the Western nations about the forests and forest products industries of Mainland China than about any other region of the world. The material which does exist indicates that, currently, Mainland China is essentially self-sufficient at a very low level of forest products consumption and it is not an appreciable factor in international forest products trade. Further, the material shows that if the region becomes an important factor in such international trade in the future, it will be primarily as a major importing area.

The most comprehensive recent analysis of the Chinese situation in forestry and forest products has been made by Dr. S. D. Richardson, Director of Research, New Zealand Forest Service, on the basis of a review of the various scattered published materials bearing on China, and a study tour of six weeks in the region [ref. 20]. Another analysis has been published by Mr. Jan J. Solecki of the University of British Columbia, on the basis of a literature survey, including a substantial amount of material published in the Soviet Union [ref. 21].

The FAO [ref. 11] has estimated that Mainland China has about 23 per cent of the world's population but only 2.5 per cent of the world's accessible forest area. Dr. Richardson accepts a total forest area of 237 million acres and a total volume of 265 billion cubic feet as the best available estimates of the region's forest, and quotes Chinese sources claiming that 75 per cent of this area will ultimately be accessible. The majority of China's forests are located in the northeast, while the main areas of population concentration and industrial activity have little or no timber.

Dr. Richardson also cites a 1958 analysis made in China indicating that by 1972 the requirements for industrial wood were expected to reach 4.2 billion cubic feet, while planned output was 2.8 billion cubic feet, leaving a gap of 1.4 billion cubic feet which might be taken as some sort of an indication of possible future import levels. Realization of any such import levels would, of course, depend upon developments in the economy and forest industries of the region along the lines envisaged in this analysis, plus the development of a substantial volume of foreign exchange by China and the decision to allocate part of such foreign exchange to wood imports.

To the extent that import markets do develop in China, Southeast Asia may prove to be a source of sawlog-size material. For both pulp and softwood timber, however, the two possible sources appear to be Siberia and North America (particularly Canada). In such a trade, North American sources might not be at a disadvantage. Solecki summarizes this question as follows:

> At first sight the Soviet Union appears to have an advantage, but this is not necessarily so. The timber stand of Eastern Siberia is needed for the processing centres recently completed or under construction on the Angara and the Yenisei rivers. In addition the cost of transporting timber to the North Sea and transloading it to ocean going ships in the short ice-free season is likely to be rather high, while transport by rail would be even more so. The forest stands of the Far East (Maritime Province) on the other hand are primarily still underdeveloped mixed forests with very high costs of exploitation. Their proximity to the Chinese markets is also somewhat illusory. The Amur flows East and then North, where the handling of timber becomes difficult due to prolonged winters. Transport by rail, or small craft by water against the current, will be expensive.

Obviously, any speculations as to future export markets in Mainland China for North American timber are subject to major political uncertainties as well as to a high degree of economic uncertainty. To date, no such markets have emerged on an appreciable scale, even though in the pre-World War II period China relied heavily on imports to meet its forest products requirements.

The U.S.S.R.

The U.S.S.R. has the world's most extensive softwood forest reserves. Lumber production has risen rapidly over the last decade, and by 1963 the nation produced 30 per cent of the world's total output of softwood lumber. With this expansion of production, the U.S.S.R. has resumed its pre-World War II role as one of the main softwood lumber exporters of the world. The pulp and paper industry is much less developed, but a very great expansion is planned for the future.

In 1963, the total value of forest products exports from the U.S.S.R. was $409 million, while forest products imports totaled only $87 million. Lumber and wood exports represented 84 per cent of the export value, but the nation also had export surpluses in pulp and in wood manufactures. Only in the paper and paperboard classification was the value of imports greater than the value of exports.

The U.S.S.R. very clearly is not going to be a source of demand upon or opportunity for forest products from North America, but instead will be an important source of competition with North American forest products in world markets. Accordingly, the main discussion of the Soviet outlook in forestry and forest products is presented in the following chapter on supply possibilities.

Australia and New Zealand

Australia and New Zealand both have well-developed economies and high per capita levels of forest products consumption. However, with 10.7 million people in Australia and 2.5 million in New Zealand, their combined impact on world markets in forest products is relatively minor.

The native forests of Australia are largely hardwood types, and, despite some development of softwood plantations, the nation remains strongly dependent on imports for softwood forest products. In 1963, forest products imports totaled $146 million, compared to $11 million in exports. About 60 per cent of the import value was in paper and paperboard, with softwood lumber and wood pulp being the other main import groups. In New Zealand, the great development of softwood plantations enabled exports of $30 million, which gave a slight surplus over forest products imports of $28 million. Paper and paperboard represented more than half of both the imports and the exports. The nation also exported softwood logs and lumber and some wood pulp, while importing higher grades of softwood lumber and some wood pulp.

The FAO [ref. 6] estimated that net imports by the two nations combined represented a roundwood equivalent of 0.03 billion cubic feet in 1961, and projected that this excess of consumption over production might rise to 0.16 billion cubic feet by 1975.

Exports from Canada to Australia in 1963 included 118 million board feet of lumber, 2 million square feet of softwood plywood, 89 thousand tons of newsprint, 9 thousand tons of other paper and board, and 40 thousand tons of wood pulp. Exports from the United States to Oceania generally (primarily to Australia) included 108 million board feet of lumber in 1963, and 22 thousand tons of pulp plus 30 thousand tons of paper and paperboard in 1962.

The Developing Nations

This grouping of all the other nations of the world under a single heading for relatively brief treatment may seem to be a somewhat cavalier process. It is justified, however, by the relatively minor interrelationships between the U.S. forest resource and these extensive areas of the world.

In an interesting study of the possibilities for expanding forest products exports from the developing nations, the FAO [ref. 12] grouped the nations of the world under three headings: the developed countries (the United States and Canada, the western and southern European nations, Japan, Australia, New Zealand, and South Africa); the centrally planned economies (the U.S.S.R. and the eastern European nations); and the developing countries. The analysis of the aggregate trade flows in forest products during the period 1959–61 showed that 71 per cent of the total value of the trade was in trade between developed countries; that 10 per cent was from developed to developing countries; that 6 per cent was from developing to developed; and that the remaining 13 per cent did not involve the developed countries.

In summarizing the prospects for an expanded export trade from the developing nations, the study states: "Western Europe is destined to remain the principal prospective market for the developing countries. . . . The next most important market is likely to be Japan, a rapidly expanding, high wood-using, economy which is progressively running into wood deficit." In North America, the main opportunity for expansion foreseen is in hardwood plywood and veneer.

These conclusions generally parallel the findings of the present study, in which it has been argued that Europe and Japan are the primary present and potential sources of external demand on the North American forest resource and that North America is likely to be self-sufficient in its forest products consumption patterns except for the less than 1 per cent of consumption represented by tropical hardwoods.

The FAO [ref. 6, 7] has published detailed analyses of timber trends in the Asia-Pacific region and in Latin America; by the spring of 1965 a similar study in Africa had been brought to a completed manuscript stage. On the basis of these and several related studies, the FAO has estimated the total consumption of forest products in the developing countries in 1959–61 and has projected these consumption levels to 1975 [ref. 13]. The summarized estimates are presented in Table 7.

The total annual consumption of forest products in the developing nations of the world during 1959–61 is estimated at 3.88 billion cubic feet of industrial roundwood equivalent. The projected level for 1975 is 6.50 billion cubic feet, giving a total increase over the fifteen-year period of 2.6 billion cubic feet.

These estimates, of course, do not show the industrial roundwood volume moving out of and into the developing countries through international trade with the developed and centrally planned countries. The FAO analysis [ref. 12] shows that the total value of forest products imports into the developing countries from these outside sources was $558 million, while the total volume of exports to the other nations was $349 million. No data are given as to the industrial roundwood equivalents of these import and export values.

Exports from North America to the developing nations in 1963 represented a roundwood equivalent of about 0.17 billion cubic feet, or about 16 per cent of the total wood volume represented by all exports. Roughly 55 per cent of the export volume originated in the United States, and 45 per cent in Canada. Softwood lumber accounted for nearly one-third of the export volume, with the balance being primarily wood pulp, newsprint, and other paper and board. About three-fifths of the softwood lumber exports came from the United States, as well as a little more than half of the wood pulp and most of the paper and board other than newsprint.

Imports of forest products based on wood originating in the developing countries was about equal to or slightly less than this volume of exports. These imports consisted primarily of hardwood plywood and veneer (much of which was manufactured in and imported from Japan) plus small volumes of hardwood logs and lumber.

If the total consumption of industrial forest products in the developing nations were to increase to the full amount projected by the FAO by 1975, the extent to which this would generate additional demand for imports from the developed and centrally planned nations would, of course, depend on the ability of the developing nations to expand their own production. In assessing

TABLE 7. Consumption of Wood in Developing Countries, 1959–61, and Estimated Requirements in 1975

Region	Lumber		Plywood		Fiberboard and particle board		Paper and paperboard		Roundwood		Total roundwood equivalent	
	1959–61	1975	1959–61	1975	1959–61	1975	1959–61	1975	1959–61	1975	1959–61	1975
	billion cubic feet				*million tons*				*billion cubic feet*			
Latin America	0.53	0.82	0.009	0.014	0.15	1.00	2.88	7.34	0.30	0.37	1.65	2.76
Near East	0.05	0.07	0.003	0.009	0.04	0.17	0.17	0.44	0.09	0.12	0.19	0.31
Asia-Pacific	0.35	0.58	0.015	0.035	0.06	0.33	1.39	4.21	0.54	0.71	1.41	2.35
Africa	0.10	0.20	0.005	0.012	0.07	0.23	0.55	1.49	0.36	0.49	0.62	1.07
Total	1.04	1.68	0.031	0.070	0.32	1.74	4.99	13.47	1.29	1.69	3.88	6.50
Roundwood equivalent, billion cubic feet	2.06	3.32	0.07	0.16	0.02	0.11	0.45	1.21	1.29	1.69	3.88	6.50
Value, million U.S. dollars	1,340	2,160	150	330	35	195	1,040	2,815	255	335	2,820	5,840

Source: Food and Agriculture Organization, "Fast-Growing Tree Species for Industrial Plantations in Developing Countries." *Unasylva*, 19(4):159–67.

the export potential of the developing nations, the FAO study assumes that the developing nations would "cover their own extra needs of all categories save pulp and paper, of which they would import one-quarter of their extra needs, processing themselves the other three-quarters. . . ." Under this assumption, the increased consumption levels could be achieved with increased imports of only 2.1 million tons of pulp, paper, and board, representing roughly 0.2 billion cubic feet of roundwood equivalent.

The suggestion that the developing nations can achieve the degree of expansion in forest products output involved in the FAO assumption may well be considered optimistic. However, the same could be said of the projections of economic growth rates and total forest products consumption shown in the FAO studies. Certainly any projections of developmental prospects and raw material use in the developing nations must be recognized as involving a high speculative element. The FAO studies are intended only to indicate the general direction of movement and something of the general magnitudes which might be involved.

Construction methods are adaptable to indigenous materials. Major increases in softwood lumber imports by the developing nations are not likely in the context of limited foreign exchange, alternative materials, and forest resource availability which characterize the prospects for the developing nations. Pulp and paper can be manufactured from a fairly wide range of materials in addition to softwood pulpwood. Although the quality of the

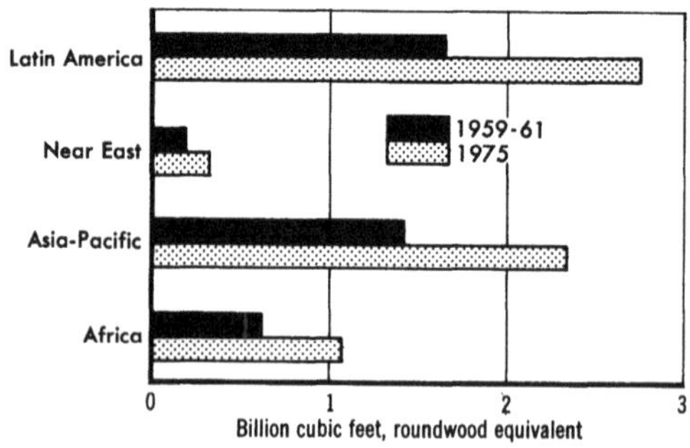

FIGURE 13. *Annual consumption of industrial forest products in the developing regions of the world, 1959–61 average, and FAO projection for 1975.* (Source: *Food and Agriculture Organization. "Fast-Growing Tree Species for Industrial Plantations in Developing Countries," in* Unasylva, *19[4]:159–67.*)

product may be less satisfactory than that of the output of North America and Northern Europe, the developing nations must be expected to develop their own production in this field, relying on some softwood pulpwood from plantations plus some expansion in imports to meet the more exacting quality requirements.

Thus, the pulp and paper field is the area in which prospects for expanded exports to the developing countries are the greatest, but even here the total volumes involved are limited as suggested by the FAO analysis. The North American share in this export market may tend to rise somewhat over the next decade as expanded European consumption acts to reduce the export volume from Northern Europe moving to these other regions. Some increase in exports from the U.S.S.R. to the developing nations may work in the opposite direction.

If it were to be assumed that the full amount of the 2 million ton import increase suggested by the FAO analysis would result from increases in North American exports, then the 1975 level of such exports would be about three times that of 1963. The *additional* roundwood equivalent would represent less than 5 per cent of the total North American cut for pulp and paper production in 1962.

CHAPTER 7

SOME ASPECTS OF THE SUPPLY OUTLOOK

There remains the question of the degree to which supply prospects in the rest of the world may have some impact on the forest resources of the United States. In view of the analysis presented in the previous chapters, the key question here appears to be the degree to which expanded production in other parts of the world will serve as a source of competition for and reduction in demands on exports of forest products from North America to the rest of the world. Behind this question there is the even broader one of the extent to which the forests of the rest of the world might become a source of supply to North American consumers.

Estimates compiled by the FAO [ref. 9] indicate that the total removal of roundwood volume from the forests of the world amounted to 64 billion cubic feet in 1963. Of this volume, 58 per cent was cut for industrial use, while the other 42 per cent was taken for fuelwood, including wood for charcoal. For the purpose of the present study, primary interest centers on the industrial wood component. Of this 36 billion cubic feet of industrial roundwood, 75 per cent was from softwood species and 25 per cent from hardwoods. The dominant role of softwoods in the world's use of wood in forest products is clearly shown. Of the removals of softwood for industrial use, 38 per cent came from North America, 30 per cent from the U.S.S.R., 21 per cent from Europe, 8 per cent from Asia (primarily Japan and Mainland China), 2 per cent from Latin America, and the remaining 1 per cent primarily from New Zealand.

As discussed in Chapter 5, North America's role in world trade in forest products involves substantial exports based largely on softwoods and a very much smaller volume of imports based primarily on tropical hardwoods. In view of the nature of this trade and the worldwide importance of softwoods in forest products use, the following discussion will place main emphasis on supply prospects for softwood production. Some attention will also be given, however, to the potential role of hardwoods, especially tropical hardwoods, relative to the timber resource of the United States.

The U.S.S.R.

From the standpoint of North American forest resources, the forests of the U.S.S.R. are of particular interest because of their vast extent, their high proportion of softwoods, and the uncertainties surrounding their potential role in forest products markets in both Europe and the western Pacific area. The most recent estimates of the FAO [ref. 16] credit the U.S.S.R. with 2,249 million acres of forest land, or 22 per cent of the estimated world total. The growing stock volume is estimated at 2,793 billion cubic feet, including 2,342 billion cubic feet of softwoods. These volumes are roughly double the corresponding figures for North America.

Great though the Soviet forest reserves are, these FAO estimates may give an exaggerated impression of their relative importance. In commenting on earlier FAO estimates covering the forests of the U.S.S.R., Professor P. V. Vasiliev indicated that the forested area actually in use was approximately 1.0 billion acres, with a growing stock volume of approximately 1,500 billion cubic feet [ref. 30].

Data presented by Perepchin and Filinov in a 1964 Soviet publication cited by Solecki indicate a lower potential than suggested by earlier data [ref. 21]. According to Solecki, these data begin with a total area under forestry jurisdiction of 2,905 million acres, of which 1,697 million acres are forested. Approximately 568 million acres are held to be unsuitable for exploitation because of economic factors related to their very northerly location. Another 136 million acres are estimated to be too steep or too marshy for exploitation. Areas currently over-exploited or restricted are estimated at 190 million acres. On this basis, the under-exploited forest area is estimated at 803 million acres.

The total volume on this area is estimated at 1,660 billion cubic feet. Assuming that this is cut over a one-hundred-year logging cycle to enable organization for continuous harvesting, the allowable cut from this area is estimated at 16.8 billion cubic feet, yielding 15.0 billion cubic feet of ready cut timber. When the 3.2 billion cubic feet of cut now established on the over-exploited areas is added, there is a total allowable cut of 18.2 billion cubic feet.

Such an allowable cut of somewhat more than 18 billion cubic feet would represent about a 44 per cent increase over the total removals of 12.7 billion cubic feet reported by FAO for the U.S.S.R. in 1963. While this would represent a very substantial increase in total output, it is far less than the doubling or tripling of cut and future cutting levels of more than 35 million cubic feet which have often been quoted without consideration of costs and accessibility.

These various estimates which have been cited are indicative of the uncertainties surrounding knowledge of the Soviet timber inventory—uncertainties which increase rapidly in magnitude when questions are raised such as geographic location, size structure, and growth rates. Of even greater significance are the importance of and uncertainties concerning the economic dimensions of this timber inventory. Accessibility, costs, and markets together will determine the extent to which the timber inventory is actually a resource to the Soviet Union. In its general elements the problem is much the same as that involved in the economic evaluation of the forests of Canada, although both the magnitudes of timber involved and the economic obstacles to its utilization appear to be greater in the U.S.S.R. than in Canada.

The central problem in Soviet forest development rests in the fact that the forests are concentrated in the north and east, while the labor force and consumption centers are in the south and west. Reflecting this situation, the historical pattern has been for a concentration of cutting in the western and southern areas, with consequent over-exploitation of the forests of these zones. In recent years, efforts have been increased to shift the logging to the heavily forested regions; appreciable progress has been made, but various resistances have been encountered, including increased transportation costs, a lack of adequately developed infrastructure, difficulties in the recruitment of labor, and increased costs related to smaller timber.

Total removals from the forests of the U.S.S.R. are reported by FAO to have increased from 9.8 billion cubic feet in 1950 to 12.7 billion cubic feet in 1963. During this period the use of wood for fuel has declined. Thus industrial wood removals have increased from 5.9 billion cubic feet to 9.3 billion cubic feet. Lumber is the major element in industrial wood. In 1963, the allocation of the industrial wood production involved 66 per cent to sawlogs, veneer logs, and logs for railroad ties; 15 per cent to pulpwood and mine timbers; and 19 per cent to other uses.

During the period 1950–60, lumber production in the U.S.S.R. rose markedly, while North American lumber production declined slightly and European production experienced only a modest increase. The Soviet share of total world softwood lumber production rose from 20 per cent to 33 per cent over this period, while the North American share declined from 43 per cent to 31 per cent. In 1963, total lumber production in the U.S.S.R. was estimated at 44.5 billion board feet, or 30 per cent of the world total; 85 per cent of the production was in softwoods, representing 33 per cent of the world's softwood lumber total.

With this increase in production, the U.S.S.R. was able both to increase its domestic consumption and to become a major factor in international trade in lumber. FAO estimates for 1960 show a per capita consumption of lumber in the U.S.S.R. of 207 board feet, which can be compared with 206 board feet for North America and a world average of 49 board feet. During recent years in North America, of course, there has been a strong swing from lumber to plywood. Per capita consumption of plywood in North America is nearly eight times as high as in the Soviet Union.

The extensive reconstruction program required in the U.S.S.R. following World War II was a stimulus to rapid development of the lumber industry. By 1950, production had made a strong recovery from the wartime low, but was not adequate to maintain a net export trade. During 1949–51, softwood lumber exports to Europe (primarily the United Kingdom) amounted to about 400 million board feet, but were almost exactly offset by imports (including reparations) in the same amount from Rumania, East Germany, and Finland. By 1959–61, however, exports to Europe had risen to 1.8 billion board feet, while imports had shrunk to a minor amount.

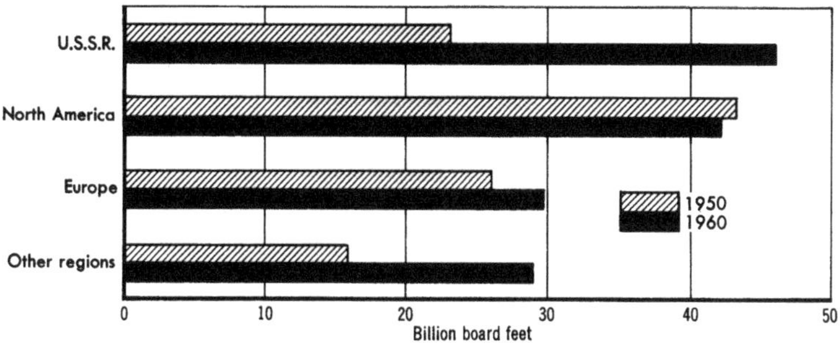

FIGURE 14. *World lumber production by major producing regions, 1950 and 1960.* (Source: *Food and Agriculture Organization.* Yearbook of Forest Products Statistics, 1964 [*Rome, 1964*].)

In 1963, the total softwood lumber exports from the U.S.S.R. amounted to more than 2.7 billion board feet. While this was very much less than the 6.3 billion board feet exported from Canada, the major part of the Canadian exports went to the United States. Thus, in terms of world markets outside of North America, the Soviet lumber exports were greater in volume than the 2.3 billion board feet of exports originating from North America. The distribution of the softwood lumber exports from the two regions differed substantially, but the European market—particularly the United Kingdom—is the leading export market for both, taking 90 per cent of the exports from the U.S.S.R. and 50 per cent of the exports from North America.

In 1963, softwood lumber imports by the United Kingdom totaled 3.4 billion board feet, of which 26 per cent came from the U.S.S.R., 24 per cent from Finland, 20 per cent from Canada, 16 per cent from Sweden, and the remaining 14 per cent from other sources.

Discussions with representatives of the lumber export trade in Sweden and Finland indicated that, although the U.S.S.R. was once considered erratic in its deliveries and somewhat disruptive in its pricing policies, it is now considered an effective marketing force with good performance and skillful pricing.

In terms of total wood volume, roundwood equivalent, the exports from the U.S.S.R. in 1959–61 amounted to 0.43 billion cubic feet, of which 60 per cent was in lumber, 28 per cent in logs and bolts, 9 per cent in wood pulp, 1 per cent in paper, and 2 per cent in wood-based panel products. With the shifting of logging toward more remote regions and the further development of domestic forest industries, expansion of the log and bolt exports is not considered likely. Some further expansion of the softwood lumber exports may occur, but not at the rapid rate of the last decade. Perhaps the greatest uncertainty is in the area of the future level of wood pulp exports.

Although the lumber industry of the U.S.S.R. is of major proportions, the pulp and paper industry has had only limited development. Wood pulp production for 1963 is estimated at 4.3 million tons, or only slightly more than 10 per cent of the North American production. Paper and paperboard is even less developed, with a 1963 output of 4.2 million tons or 9 per cent of North American production.

As these production figures would suggest, the per capita consumption of paper and paperboard in the U.S.S.R. is very low. Estimates of the FAO [ref. 15] indicate a per capita consumption of 37.5 pounds of paper and paperboard in 1963. This can be compared to 121 pounds per capita for all of Europe or to 417 pounds per capita for North America in the same year.

For a number of years, Soviet planners have stressed the need for further processing of the timber harvest, including the establishment of goals calling for rapid expansion in pulp and paper. Such a development is sought through establishment of a number of very large-scale integrated combines. Achievement, however, has fallen far short of plans. This is shown clearly in a recent compilation by the FAO [ref. 14] of various materials related to pulp and paper trends in the Soviet Union.

For example, targets for the 1959–65 Seven Year Plan at one stage included 1965 outputs of 8.8 million tons of pulp, 4.6 million tons of paper, and 4.0 million tons of paperboard. In 1963, these were cut back sharply to 4.1 million tons of pulp, 3.8 million tons of paper, and 1.7 million tons of paperboard. The problems encountered in attempting to achieve the planned expansion included a lagging of funds for investment below the levels planned, the necessity of procuring much of the equipment abroad because of the limited domestic pulp and paper equipment industry, a shortage of qualified technicians, and a lack of realistic planning for transportation and other elements of infrastructure.

However, the targets initially set in the 1964–70 Seven Year Plan were again ambitious, including a 1970 level of 12.6 million tons of pulp, 7.7 million tons of paper, and 6.6 million tons of paperboard. Much of the new production is planned for combines, located mainly in the northern European portion of the U.S.S.R., in the Ural Mountains, and in central and eastern Siberia. Achievement of these goals would require a 1970 pulp production 3 times that achieved in 1963, while paper production would be 2.4 times and paperboard 6 times the 1963 level.

Announcement of these goals was received with considerable skepticism outside of the Soviet Union, reflecting the view that factors such as the recent increased emphasis on agriculture would shift investment away from the forestry sector and make the goals greatly overstated. For whatever reason, the Soviet planners subsequently revised and drastically reduced these goals, and in February 1966, *Pravda*[1] announced the following new target for 1970: pulp production, 9.3 to 9.9 million tons; paper, 5.5 to 5.8 million tons; and paperboard, 4.6 to 5.0 million tons.

While it seems obvious that rising standards in the U.S.S.R. could absorb the planned increase in output over the period, it is certainly possible that a substantial part of any increase could be allocated to the export trade. The considerations involved in that decision are both political and economic; however, costs could present some formidable barriers to such an expansion. Much of the expansion is scheduled for Siberia, with central Siberia being the focal point for the planned combines, although a substantial expansion is planned for the northern European zone and the Urals. Shipment of pulp (or paper) from central Siberia to the European markets by rail would involve very high costs, but the only apparent alternative is development of water transportation down the Yenisei River through the Arctic Ocean and out to the Atlantic. The costs and very short shipping season of this alternative appear adverse to any such development. Nonetheless, it has been suggested that completion of the combine at Bratsk and additional combines in central Siberia would provide pulp for European Russia, thus releasing pulp from this latter zone for export to the European market.

Perhaps a more probable zone for increased pulp exports would be from developments in eastern Siberia into the western Pacific area. The U.S.S.R. has proposed development of four

[1] As reported to the writer in correspondence with members of the Food and Agriculture Organization.

combines in the Far East in addition to one already under development at Komsomolsk. It is reported that the Soviet Union has approached the Japanese government with regard to this plan, proposing that Japan supply the needed pulp and paper machinery for the projects with the payment schedules to be tied to pulp exports to Japan.

Various reports indicate that the U.S.S.R. has greater ability to manufacture pulp machinery than paper machinery. Furthermore, a considerable part of the output of paper machinery apparently will be needed to replace obsolete equipment in existing capacity. This situation, plus the technological logic of initiating pulp and paper expansion in the pulp sector suggests that the pulp phase of the intended expansion might be expected to develop more rapidly than the paper and paperboard phase, thus giving an expanded surplus of pulp for the export markets during the period in which the capacity in paper and board was being brought up to an equivalent level. This is consistent with the various targets for 1970, which appear to enable an increase in pulp exports into the general range of 1.5 to 2.0 million tons by 1970. The major part of any pulp surplus realized is planned for export to the centrally-planned nations of eastern Europe.

Certainly there is evidence that the U.S.S.R. is actively interested in expanding exports. For example, in February 1964, then-Premier Khrushchev stated that, with adequate development, exports of the wood processing industries could reach more than $500 million annually. This can be compared with a 1962 level of such exports of about $200 million to hard currency areas.

Essentially, the outlook for the U.S.S.R. becoming an increasingly important factor in world pulp and paper markets involves two major kinds of uncertainties: the degree to which planned expansions in output will actually be achieved, and the extent to which any expanded output is allocated to the export market. The longer-term potential of the Soviet forests is even more uncertain. As has been noted earlier, the most recent data from Soviet publications cast considerable doubt on the common view that, with the development of access, the forests of the U.S.S.R. could sustain a cut double or even triple current levels. These recent data suggest an allowable cutting level of only about half the frequently indicated 35 billion cubic feet. However, since all reports indicate that both logging and milling in the U.S.S.R. are characterized by a very high proportion of unused residues, a major expansion in the volume and value of the forest products produced should be possible with only a modest increase in cut.

The remoteness of location, the lack of access, and the growth characteristics of a major part of the unexploited Soviet forests appear to limit greatly their economic significance. Even with the development of access, transportation would represent a major cost item in reaching the important world markets. Furthermore, expanded domestic consumption seems likely to press hard on expanded production.

No available studies compare projected internal consumption levels with production possibilities. However, at least one view of the prospects has been concisely summarized by P. V. Vasiliev, who has said that in twenty years industrial production will increase not less than six times. Even if the norms of utilization of wood in mining, construction, railroad transport, etc., are lowered by 1.5 to 2 times, Vasiliev feels that the national economy will need a yearly supply of more than 35 billion cubic feet, which exceeds the natural productivity of the forests.

Thus, the available evidence suggests that for the next few years the U.S.S.R. will continue to face a tight wood supply and relatively high-cost production, while over the longer period rising domestic consumption will tend to keep up with expanded output. This does not preclude the possibility that, in particular developmental stages or particular market situations, the U.S.S.R. will be a very active market force. This is currently the case in the European softwood lumber trade, and a substantial expansion of Soviet pulp exports within ten years is certainly conceivable. However, it does indicate that the U.S.S.R. is not likely to be an ever-increasing source of major surpluses of wood products in world markets as has sometimes been suggested.

Northern Europe

The primary review of the outlook for the entire European region, with an emphasis on the demand aspects, has been presented in Chapter 6. However, some separate consideration of northern Europe from the supply standpoint is appropriate at this point. In 1962, Norway, Sweden, and Finland together produced 45 per cent of the total softwood timber removals in Europe. Sweden and Finland rank second and third among the nations of the world in the value of their forest products exports; the combined exports of the two nations have a value slightly higher than that of Canada. In general, northern Europe is the main surplus region in forest products in Europe and the primary source of imports of forest products by the balance of Europe. In addition, these nations maintain an active, if limited, trade in forest products on a worldwide basis.

In 1949–51, the four northern European nations (Denmark, Norway, Sweden, and Finland) produced more than 2.15 billion cubic feet of industrial roundwood, representing 35 per cent of the total output for Europe. By 1959–61 this had risen to 2.82 billion cubic feet, or 38 per cent of the European total. Wood use has shifted steadily toward the pulp and paper industry, which took 1.45 billion cubic feet in 1959–61 compared to 1.20 billion cubic feet for lumber.

Finland and Sweden are the major producers among the northern European nations, together producing about 90 per cent of the total wood removals. In Finland, removals in 1960 were believed to be slightly in excess of growth. Although silviculturally a long-term increase in cut is considered feasible, and a strong effort to increase growth is being made, available studies indicate that at best the cut is not likely to rise over the next 15 to 20 years; in fact, a slight decline from 1959–61 to 1975 is possible. In contrast, Swedish studies show that the cut can be increased by more than 20 per cent over this period. Norway and Denmark also have some possibilities for expanded cutting. Current efforts to expand forest growth in this region are centered in such intensive practices as swamp drainage, fertilization, and genetic improvement of planting stock.

Although the supply situation in Finland for timber for current cutting is particularly tight, the nation is a net exporter of unprocessed roundwood. According to FAO data, Finland imported about 0.03 billion cubic feet of pulpwood and softwood logs in 1963, but exported nearly 0.05 billion cubic feet of such material plus a substantial volume of round mine timbers, poles, piling, and posts. Sweden is also a net exporter of roundwood, primarily pulpwood. In contrast, Norway was a net importer of more than 0.03 billion cubic feet of pulpwood in 1963. These roundwood exports generally represent movements across adjacent borders in response to transportation considerations.

Tending to offset its wood supply problems, Finland has the most modern and efficient mill capacity in pulp and paper of the Northern European nations, reflecting the aggressive development of the industry during the last decade. Swedish mills on the average are both older and smaller than Finnish mills, while Norwegian mills in turn tend to be older and smaller than those of Sweden.

For all four northern European nations, the FAO indicates an increase in total removals of about 10 per cent may be possible by 1975 [ref. 10]. Such an increase would be concentrated in hardwoods (with a 30 per cent increase) and pine (with a 12.5 per cent increase), while a decline of 4 per cent is shown for the cutting level of spruce. With a further shift from fuelwood to industrial wood, total industrial wood removals might rise from 2.7 billion cubic feet of softwoods in 1959–61 to nearly 3.1 billion cubic feet in 1975, while hardwood removals might increase from 0.15 billion cubic feet to 0.42 billion cubic feet over the same period.

Essentially, further expansion of output from northern Europe appears to be sharply constrained by the availability of wood, with any further increase in cut involving problems of rising costs and a shift to lower-valued species. Against this tight wood situation, efforts to expand the value of forest products output can be expected to lead to a continued upgrading of products, with a further shift from lumber to pulp and further vertical integration from pulp into paper and paper products. The extent of such upgrading in the pulp and paper field will, of course, be strongly affected by developments within the European Economic Community.

Even with the projected increases in cut and the shift in product mix indicated, northern Europe appears to be far from able to supply the expanding consumption of forest products in Europe. Indeed, nearly a third of the projected expansion is likely to be taken by rising domestic use within the northern European region.

Fast-growing Plantations

The softwood forests of the world are largely concentrated in the cool temperature zones of the northern hemisphere, so that North America, Europe, and the U.S.S.R. together account for 92 per cent of all softwood forests currently in use.

Although softwood forest types do occur naturally in other regions of the world, such forests are usually a relatively small portion of the total forest area of these regions, are heavily used in some instances, and often tend to occur only in scattered areas, frequently difficult of access. In Central America the softwood types are closely related to North American forest types; they are limited in total extent but are still in the process of development. In South America, the Paraná pine stands of Brazil and, to a lesser extent, Argentina, are generally under heavy use at present. In Asia scattered stands of pine occur in India, Burma, Thailand, Cambodia, Viet-Nam, Sumatra, Java, Borneo, and the Philippines, occurring generally in hilly regions. While locally important, their total volume is small. The more important softwoods of Asia are in temperate zone forests in Japan, Korea, Mainland China, and the western and central Himalayas. The accessible portions of these forests are already under heavy use. In Africa, native softwoods are limited to Ethiopia, plus small amounts in a few other areas.

Under the influence of this limited range of softwood forests in extensive regions of the world and the importance of softwoods to the forest industries, there has been considerable interest in the development of fast-growing plantations of softwoods. The most notable success in such efforts has been with Monterey pine (*Pinus radiata*) in New Zealand, South Africa, and Chile.

Such fast-growing plantations are one of the primary areas of interest in the forestry program of the FAO [ref. 12], which contends that the high rates of growth possible can be a "decisive advantage" to some of the developing nations in the softwood forest products industries, stating as follows:

> In Chile, Brazil, and parts of East and Central Africa, coniferous fibre is being grown at 5 to 10 times the rate (per hectare per year) that is possible in the cool temperature zone of the northern hemisphere where pulp and paper production today is heavily concentrated. And there are many other areas, in Africa, Latin America and Asia, where new resources—in the shape of plantations of quick-growing conifers—could be created in suitable locations.

The interest in fast-growing plantations is by no means limited to conifers, and, in fact, in the developing nations hardwood plantations are much more extensive than softwood. A recent survey of fast-growing plantations in the developing nations by the FAO [ref. 13] showed 2.2 million acres of softwood plantations and 8.6 million acres of hardwoods. The plantations of South Africa and New Zealand are not included in this compilation, since these areas are classed among the developed nations.

The New Zealand plantations are concentrated in a limited range of age classes which are now maturing. The quality of much of the material is low and the nation currently faces something of a problem in marketing its surplus softwoods. Chile also has a substantial softwood growth surplus and is currently attempting to develop a pulp and paper industry oriented to export markets as a means of utilizing this material. South Africa is also developing pulp and paper production, and some export surpluses.

As these examples indicate, fast-growing plantations can contribute to international trade in softwoods. However, the total area of such exotic softwood plantations is only about 4 million acres. While they can be of importance in particular local situations in international trade, their total volume is too limited to be of appreciable significance to evaluations of and planning for the North American forest resource.

The Food and Agriculture Organization estimates the current rate of establishment of fast-growing plantations in developing nations, both hardwoods and softwoods, at about 0.8 million acres per year. Even with an optimistic estimate of the yield of these plantations, the conclusion was that their output would not keep up with rising consumption levels in these areas of the world. Unless the programs are expanded very rapidly in the near future, it seems unlikely that such plantations will make any very appreciable contribution to forest products markets in the developed economics for many years to come.

Tropical Hardwoods

The forest areas of Africa, Asia, and Latin America are reported to total 5,300 million acres, representing 54 per cent of the total forest area of the world [ref. 16]. Less than one-quarter of this vast area is currently in use, while more than half is classified as inaccessible. A very wide range of forest types is represented in this total area, including a limited proportion of softwood forests, a substantial proportion of temperate zone hardwood and mixed forests, and a very great area of tropical forests, predominantly hardwood in nature and including both evergreen and deciduous types.

The softwood types and the temperate zone forests generally have a fairly clear role in world wood supplies. As access is developed, these forests can be expected to be brought into use with existing technologies, largely to meet relatively local demands. Indeed, a substantial part of such forests is already receiving heavy use. Only limited and temporary surpluses for export to international markets lying beyond the immediate neighboring countries are to be expected.

Far greater uncertainty must be attached to the potential role of the extensive tropical forests of these regions. In large part, these great reserves of tropical hardwoods remain with little or no industrial development even today. In any evaluation of North American or European timber resources, the issue of potential significance of these tropical forests presents itself.

Obviously, a very wide range of conditions is encountered in these great areas of tropical forests extending over three continents. Nevertheless, some broad generalizations can be made. Characteristically, such forests are overly rich in species, with hundreds and sometimes even thousands of species being found within a relatively limited range. Further, the distribution of trees of any one species is usually scattered, with a typical density being only one or two trees per acre of a merchantable size and a particular species. Commercial timber operations have tended to be concentrated in the limited instances where higher densities are found, such as in teak and lauan. Along with this abundance of species there is characteristically little or no knowledge of the wood properties and a lack of experience with utilization requirements. Many of the forests have been degraded by fire and shifting cultivation. Access is commonly lacking, and is difficult and expensive to develop. Distances from the main population and consumption centers are often great. In many instances the volumes of bole wood per acre are quite low.

In short, the lack of development of these forests is the result of very great economic barriers to such development, and as a concomitant of this lack of development there is a stultifying lack of knowledge of the kinds of wood represented. Far from being forest products exporting areas, the tropical forest nations of the world are typically net importers of forest products, exporting primarily various types of specialty logs. And this net import basis persists despite extremely low per capita consumption levels of manufactured forest products.

The current export pattern is largely concentrated in the veneer and plywood field, plus small volumes of woods of very unusual properties such as balsa. The main flow to North America is lauan, primarily from the Philippines and secondarily from North Borneo, shipped after processing (largely in Japan) into veneer and plywood. In contrast, Europe imports wood in log form from Africa for manufacture into plywood and veneer. The volume of these imports has reached a level at which it has adversely affected the marketability of the domestic beech, which has been a customary source of utility veneer and plywood. In addition, Europe imports teak and some other specialty hardwoods from Asia. Only a small volume of South American woods are exported.

As has been discussed earlier, such tropical hardwood imports constitute less than 1 per cent of North American industrial wood use. The greater part of this use is for decorative effects and no great problem would be involved in substituting alternate materials for the tropical plywoods and veneer. While the ability of current sources to maintain their flow of exports at current or expanded levels for a long period may be questionable, extensive additional sources of tropical hardwood plywood logs appear likely to be developed from still untapped tropical forests.

In view of the rapid increase in consumption levels of pulp and paper in all the developed regions of the world, and of the great importance of pulp and paper to the outlook for export demand on North American timber, particular attention must be given to the potential role of tropical hardwoods as a source of pulp and paper for the world. This question was discussed with a wide range of experienced individuals during the course of this study, and in no instance was strong support expressed for the future role of tropical hardwoods in the pulp and paper industry.

The basic question is not one of technological feasibility. Pulp can be and has been made from a number of tropical hardwoods. If sufficiently large volumes of wood are processed to enable sorting of the wide mixture of tropical logs by fiber properties, there appears to be little reason to question the technological feasibility of pulp mill operations based on tropical hardwoods.

The real issue is that of economic feasibility. As a result of their nature and their location, natural tropical hardwood forests do not in general appear to be a cheap source of fiber. Studies conducted in Asia, in Africa, and in South America have indicated that fast-growing plantations in favorable locations and a wide range of nonwood sources of fiber will ordinarily provide cheaper sources of fiber than tropical hardwoods. The advantage lies with the natural tropical forest only in a few unusual cases. This is clearly reflected in both the currently established and planned pulp capacity in the regions with tropical forests.

For example, on the basis of a detailed study of pulp and paper prospects in Latin America, the FAO [ref. 8] concluded that tropical forests were not likely to be a substantial source of pulpwood by 1975. Instead, the main sources of wood were shown to be the plantation-grown pine in Chile, the pine forests in Central America and Mexico, the pine forests and eucalyptus plantations in southern Brazil, and the poplar and willow plantations in the Paraná Delta in Argentina. The study also indicated that sugar cane bagasse is likely to provide an increasingly large proportion of the industry's fiber supply.

Similarly, the limited pulp industry of Africa is based entirely on exotic plantations and nonwood fibrous materials. The main center of production is in South Africa, which has some 2.5 million acres of plantations, with about 40 per cent in softwoods and 60 per cent in hardwoods. Plantations in East Africa are considered particularly promising for further expansion of the industry.

Much the same situation exists in Asia, with primary reliance on the limited softwoods, bagasse, and bamboos, plus substantial emphasis on plantations in plans for the future.

Some limited development of the pulp and paper industry on the basis of tropical hardwoods is definitely in prospect, but only in special situations and for local markets. The tropical forests must be recognized as part of the world's ultimate fiber potential, but it is a potential that is not likely to be realized in the pulp and paper field unless the costs of pulpwood from temperate-zone forests, fast-growing plantations, and pulping fibers from nonwood sources rise very sharply. In a world of exploding populations and hoped-for increases in living standards, such a rise in the cost of the established pulping materials may emerge and tropical forests may become an important source of fiber. At present, however, any such development still seems remote.

CHAPTER 8

AN EVALUATION OF WORLD TRENDS AND THEIR RESEARCH SIGNIFICANCE

In Chapter 1 the two main objectives of the study were stated as follows: (1) to identify the patterns and expectations in wood availability and wood consumption throughout the world which may impinge on production policies and public policies affecting timber in the United States; and (2) to formulate recommendations for a research program in international timber developments which can provide information useful in guiding timber production decisions and policies in the United States.

The discussion thus far has been directed to meeting the first objective. This chapter aims at meeting the second. First, an effort is made to pull the threads of the argument together in a summary of the major patterns and trends; second, the implications of these findings to research needs are developed.

North American Timber in a Developing World

During the last twenty-five years both the growth of the forests of the United States and the degree of dependency of U.S. forest products consumers on imports have increased substantially. By 1962, the cubic-foot growth of all timber in the nation was 60 per cent greater than the cut, yet nearly 15 per cent of the round-wood equivalent of the forest products consumed by the American people represented an excess of forest products imports over exports. The apparent anomaly in this situation is also reflected by the fact that the United States is the only one of the developed nations which has a surplus of forest growth over cut but which is not also a net exporter of forest products.

Much of this apparent anomaly is explained, however, by recognition that this import dependency in wood is derived directly from relationships with Canada. In 1963, Canada was the source of 88 per cent of the roundwood equivalent of all forest products imports by the United States. In its wood balance with the rest of the world, the United States was a net exporter of wood by a 2 to 1 ratio of the roundwood equivalent of exports to that of imports.

Correspondingly, the United States is the major market for the Canadian forest industries. In 1960, the roundwood equivalent of Canadian forest products consumed in the United States was more than half again as large as the volume consumed directly in Canada.

Canada and the United States are two nations nearly equivalent in total land area, but with a tremendous disparity in populations. They share a common democratic tradition, a common language, highly similar economic institutions, a common technology, access to common sources of investment funds, and even a substantial number of common corporations. They also have common forest associations, with the Canadian forests being the direct northern extension of the forests of the United States. From a resource standpoint, they constitute in large degree a single economic region, separated only by a political boundary and certain related barriers to trade.

If the two nations are considered as the North American region in analyses of timber resources, then the region is a net exporter of wood. Total exports of 1.1 billion cubic feet in 1963 represented 9 per cent of the cut of industrial wood in that year. These were offset in part by imports of 0.2 billion cubic feet, leaving a net export surplus of about 7 per cent of the industrial cut.

The productive forest area of North America is 3.0 times that of Europe, the volume of growing stock is 3.4 times that of Europe, yet the total cut in 1962 was only 20 per cent greater than that of the European average for 1959–61. As these rough comparisons suggest, the forests of North America have a very great potential for supporting increasing cutting levels under intensive management.

On the basis of detailed studies, the U.S. Forest Service [ref. 28] has recently projected a 95 per cent increase in the roundwood equivalent of U.S. timber products consumption from 1962 to the year 2000. This projection apparently postulates a great increase in the effectiveness of the forest product industries in product development and marketing, since from 1925 to 1962 only a 20 per cent increase was experienced. In any case, if this projection is accepted for present purposes, and if an equivalent percentage increase is assumed for Canadian consumption, the total domestic consumption of wood products and primary fuelwood in North America in the year 2000 would be less than the 1962 total of annual growth in the United States plus allowable cut in Canada, which is estimated at 27½ billion cubic feet. If, as is quite conceivable, domestic consumption expands to a lesser degree, the surplus would be greater.

The primary point to be made is that the North American timber inventory is adequate to support a major increase in domestic consumption while still having a substantial surplus to sustain expanded production for export markets. Such an expansion could not involve a simple across-the-board increase of the existing forest products. The mix of species, sizes, physical characteristics, and economic locations of the timber available for such expanded outputs differs widely from the mix represented by the current cut.

An obvious question can be raised about the relevance of much of the Canadian timber to such expansion, since as recently as 1961 some 25 per cent of the productive forest land of Canada was classified as inaccessible. A directly parallel, but less obvious, question might be raised about much of the surplus of growth over cut in the United States, for much of this represents combinations of species, sizes, and locations which are not now used. In an economic sense, such timber is just as inaccessible as some of the remote Canadian timber, for it, too, cannot now be logged and transported to markets which will yield a return greater than the costs involved.

In short, major changes must take place in processing techniques, in product development, and in marketing if the forest product industries are to expand on the timber base which exists—but with such changes much of what is now simply a physical material can become an economic resource. And from the functional concept of a resource, it follows that technology and markets are elements as important in resource evaluation as the physical base.

It is from the standpoint of market opportunities as an element in resource definition that the developing world economy is likely to be of the greatest importance. North America is essentially self-sufficient in its wood requirements with the exception of a comparatively small volume of tropical hardwood imports, and it has the timber base to remain self-sufficient through appropriate technological adjustments for at least the next few decades. The tropical hardwood imports, largely in the form of plywood and veneer, represent appreciably less than 1 per cent of total domestic consumption of wood in North America.

Although production of forest products in Europe is high and rising, the region has become a net importer of wood on balance and provides the leading source of external demand on North American timber. All available studies indicate that this import dependency will increase in the future. For example, an FAO study indicates that net imports may rise from 0.6 billion cubic feet in 1959–61 to as high as 2.5 billion cubic feet by 1975 [ref. 10]. Much of the expansion in imports shown in this and other studies of Europe is centered in the long-fiber, high-strength pulps and related grades of paper based on softwood pulpwood.

Such estimates of expanded trade opportunities are usually based on a simple gap model, with the import volume being estimated as the difference between the projected consumption level and the estimated regional production possibilities within the constraint of allowable cut standards. Various considerations indicate that expansions in output in Europe may encounter rapidly rising costs on the intensive margin. Thus, the possibility exists that economies of scale and rapid development of mechanization could give North American producers an economic advantage in the European market before European production reaches its projected maximum levels. Indeed, the current expansion of pulp exports to Europe suggests that exactly this is already taking place. Information on comparative costs in the expansion of forest products output among the major regions is extremely fragmentary, however.

The possibility also exists that North American producers will have a cost advantage over European producers in the expansion of short-fiber hardwood pulps. For example, the nature and location of the extensive hardwood forests of the Northeast may enable lower-cost development than will be true of many of the European hardwood stands. Political barriers to any such development may be encountered, however, since most European nations are strongly interested in developing their own short-fiber pulp possibilities.

Conversely, if European consumption of wood products rises less rapidly than projected or if cutting levels have been underestimated, Europe's ability to supply its own markets may be greater than indicated by the current analyses.

On the opposite side of the world, Japan is the main focal point for international trade in wood involving North America. The volume of such trade has risen very rapidly in the last ten years. With very rapid economic growth and a large population relative to the nation's land area, Japan tends to be active as an importer of raw materials and exporter of manufactured products. Imports of softwood logs from the United States and softwood lumber from Canada have increased rapidly in the last decade, with the logs representing substantially the greater part of the total volume. Softwood pulpwood and other logs are also imported from Siberia. In addition, Japan is an aggressive importer of tropical hardwood logs from the Philippines and other Asian nations, and its veneer and plywood industry provides the primary channel through which such wood is imported into North America.

In both Western Europe and the western Pacific, the main export market opportunities appear to be based on softwoods. The greatest potentials for such exports are to be found in North America and the Soviet Union. Thus, the external demands on North American timber will depend both on the magnitude of these market developments and the extent to which the U.S.S.R. participates in them.

The U.S.S.R. has the most extensive softwood forests of the world, but a substantial part of these forests are remote, difficult of access, and of questionable significance as a resource. Through the very rapid expansion of its lumber industry in recent years, the U.S.S.R. has become a major exporter of softwood lumber to Europe while also achieving a high level of domestic consumption. This achievement appears to have involved very heavy cutting in the more accessible forests. The extent to which a similar expansion will be achieved in pulp and paper in the future is one of the major uncertainties in world trade in forest products.

There are at least two major sources of uncertainty as to future exports of pulp and paper from the Soviet Union. The first is whether the nation will be more successful in the future in achieving its stated production goals in pulp and paper than has been true in the past. The second is the share of any expanded output which will be assigned to the export trade rather than to increasing the remarkably low level of domestic consumption in this area. On balance, an appreciable increase in pulp exports appears likely in some future period, at least as a developmental phase while expansion in paper and paperboard capacity lags expansion in pulp capacity. Over the long term, however, the evidence suggests that the U.S.S.R. may not prove to be a bottomless source of softwoods as has sometimes been suggested, but rather that expanding domestic requirements will press heavily on the economic production possibilities.

There is also some reason to expect that export expansions in pulp and paper may develop in the western Pacific more than in European markets. Current interest in expanding production in the extensive eastern Siberian forests which are remote from the main population centers of the U.S.S.R. will depend on export market expansion for its realization.

In terms of future forest products markets, Mainland China is an enigma. The limited information available suggests that, with economic development, the nation will face major deficits in its wood sources. Conceivably it could become an important market for Canadian pulp and an important element in the total wood trade of the western Pacific, but there is currently little or no sign of development of these possibilities.

The less-developed nations of the world do not appear to be destined to become an important source of demand for North American forest products during at least the next two or three decades. During the period 1959–61 the total flow of forest products from all the developed nations to the developing nations represented only 10 per cent of the aggregate value of international trade in forest products, and even the rather optimistic FAO projections of future forest products consumption in these nations suggest only a limited expansion of such exports. Construction methods are generally adaptable to local materials. In the allocation of the limited funds for imports held by these various nations, the importation of machinery for the manufacture of pulp and paper from local fiber sources is likely to receive higher priority than the importation of large volumes of manufactured pulp and paper.

Similarly, these developing nations do not appear likely to be an important source of wood imports for North America, with the exception of the various tropical hardwood specialties which, as noted earlier, constitute less than 1 per cent of North American wood consumption. While there are very large, undeveloped reserves of tropical hardwood forests in Asia, in Africa, and particularly in Latin America, their utilization even for local markets involves formidable problems. Such forests cannot yield general construction lumber that would be competitive with softwood lumber in the developed economies. Technologically, it appears feasible to manufacture pulp and paper from such forests, but all available evidence suggests that only in special circumstances will tropical hardwoods prove to be a source of low-cost fiber.

Fast-growing softwood plantations in some of these regions have great promise in the production of low-cost wood. However, the total area of exotic softwood plantations is only about 4 million acres and their products will obviously be of very limited significance in world terms. Generally, the current development of fast-growing plantations of hardwoods as well as softwoods is predicated on the meeting of local needs.

These, then, appear to be the very broad patterns and trends through which the timber resources of the United States will be interrelated with those of the developing world over the next ten to twenty years. Behind these operating economic forces, there are some ultimate issues which are perhaps more philosophical than economic in nature.

The per capita consumption of industrial wood in North America is approximately 7.5 times as high as in all the rest of the world. If the per capita consumption of the rest of the world were to rise to half the North American level, the total industrial cut from the world's forests would be increased 3 times for the present size of population. And if this increase in consumption levels were to take place over the next forty years, the population of the world would have doubled if the present rate of 1.8 per cent annual increase is maintained. An increase of six times in the annual cut of industrial wood from the forests of the world would then be required by the year 2005 simply to bring the average for the rest of the world to one-half the current North American standard.

Such figures, of course, are simply a restatement of the familiar theme of the vast disparity between resource consumption levels by the North American population and by much larger population blocks throughout the world. The consumption levels and cutting levels indicated are entirely hypothetical and presumably not relevant to the probable forces which will be working on North American timber resources during the next two or three decades through normal economic channels.

Similarly, it can be pointed out that China and India together have nearly two-fifths of the population of the world and more than five times the population of the United States. In each of these two great population centers of the world, the per capita consumption of pulp and paper products is less than one-half of one per cent of the per capita consumption in the United States. There is here some sense of a vast human need for pulp and paper which, in view of the limited forest inventories of the two nations, might be expected to exert strong pressures on timber resources of other regions. Yet in fact neither India nor China is included on a listing of the fifty main export markets for the pulp and paper from the United States at present, and there is no indication that any such demand on U.S. timber will be generated during the next few decades.

Thus, the present study has been limited to an attempt to identify patterns and trends of probable development which may be important in evaluating the timber resource position of the United States. The focus has been on forces that may work through what might be termed normal economic channels. In a philosophic sense, such a model may be wholly inadequate, for it gives little or no expression to the basic human dilemma now confronting the world.

These issues of world population growth and human need require a far broader canvass than the present study, and there is no reason to believe that an analysis of timber resources would provide a useful vehicle for their exploration. The problem analysis presented here has been kept in the far narrower framework of conventional economic forces. The climactic issues of exploding populations, however imminent they may be, as yet show little impact on the economic forces affecting the timber resource of the United States.

The Implications for Research

The analysis presented in this report has not been designed to develop new material or to lead to definite conclusions. Instead, it has been intended as a survey and co-ordination of existing information and points of view for the purpose of developing insights into needed areas of research. To the extent that various conclusions concerning timber resources have been implied, these should be considered simply as hypotheses for further exploration.

The main research implications which appear to follow from the analysis will be brought together through a series of short problem statements below. Attention is again invited to the central criterion for selection of these research proposals, that is, that they involve developments outside the United States which appear significant for evaluations of the timber resource position of the United States.

Timber trends in North America. The analysis has shown clearly the pronounced and growing interdependency between the United States and Canada in timber and the forest products industries. The view has been advanced that North America should be considered a single region for purposes of timber resource evaluations.

Such an analysis was proposed as a phase of the worldwide studies of timber trends conducted by the FAO. The study was to have been conducted through the co-operation of the U.S. Forest Service and the Canadian Department of Forestry, with the Canadian contribution to be based on new studies of timber trends in Canada and the contribution of the United States to be based largely on the recent timber trends report of the U.S. Forest Service [ref. 28]. Various delays, however, led to a tabling of this plan.

While such a systematic merging of studies in the two nations would have been highly useful, the need is for an analysis on a truly regional basis, with the entire program being developed under a single co-ordinated plan and leadership. Indeed, the burden of proof might well be shifted to the justification of any further attempts at comprehensive analysis of the U.S. timber resource in isolation from that of Canada. The current practice, for example, gives recognition to Alaskan timber which is substantially more remote from the main consuming centers of the United States than is the bulk of Canadian timber. Such a basis for defining the resource region appears to overweigh political ment of U.S. forest resources with only incidental attention to subdivisions relative to economic reality. To plan for the develop-Canada as a source of imports is to misidentify the actual resource base of U.S. forest products consumption and to ignore or deny the established practice of many forest products corporations.

The role of domestically oriented research. As has been stressed repeatedly here, the relationship between the United States and Canada in timber and forest products is such that the appropriate region for domestically oriented timber resource evaluations involving the United States is the North American region. If so defined, the domestic timber resource system is close to being a self-contained system. Total exports from North America amount to 9 per cent of the industrial wood cut, while net exports amount to 7 per cent. Some increase in the relative importance of exports seems likely in the future, but such export expansion is unlikely to develop to a degree resulting in resource scarcity for domestic consumers. In its very nature, such expansion of forest products exports must depend on a level of resource availability enabling low-cost expansion of output.

In terms of the components of the limited volume of imports, only the tropical hardwood products, representing less than 1 per cent of total wood consumption in North America, are necessarily obtained from outside the region.

The important conclusion from this is that evaluations of the timber resource outlook for the people of the United States can be oriented primarily to the North American region. Broader international studies will be of significance to particular phases of such evaluations and will be of interest in their own right, but they can appropriately be relegated to a secondary position in the pattern of studies.

This reflects the basic economic characteristics of timber and forest products, which lead to a high degree of resource orientation of the forest products industries, and to a strong inverse relationship between per capita consumption levels and distance from timber stands. Thus, the economic circumstances relevant to timber resource evaluations differ fundamentally from those relevant to some other resources, such as deep-sea fisheries, in which the resource is essentially an international resource, or petroleum, in which foreign sources are one of the dominant elements in domestic consumption patterns and prospects.

In the planning of domestically oriented research related to timber resources, recognition should be given to the strong parallels between export demand patterns and the areas of strength in domestic demand patterns. In particular, it is evident that the greatest opportunities for the expansion of exports are based on utilization of the softwoods. In contrast, the broad pattern of adjustment of the forest product industries to the timber available in North America must involve an increase in the relative role of the hardwoods. Export markets are not likely to be of great importance in meeting this kind of utilization problem. The solution to such timber resource problems must be sought largely within domestic markets.

North American trade barriers to efficient development of the timber resource. Since the North American region is divided into two major nations, there is an opportunity and need for research on the effects of this political division on the efficiency of timber resource use and development. Such research is a logical corollary of the regional timber trends study proposed above, but it could be developed independently of such regional studies.

Such research should be designed to determine the effects of various politically oriented trade barriers, both visible and invisible, on the nature, timing, and structure of the development of the forest products industries on both sides of the border, and through this to estimate the effects of these barriers on the efficiency of resource use. As a part of such research, analyses might be made as to the probable short- and long-term effects of modifications of these trade barriers.

Some aspects of this research area have been developed through various tariff hearings and negotiations, but these reports have generally been narrowly conceived and not directly concerned with resource issues. The Private Planning Association of Canada has recently initiated a study of the relationship of free trade and trade barriers to Canada's role in economic affairs of the North American and European regions. The study is receiving partial support from a Ford Foundation grant. The pulp and paper industry is one of the fields to be emphasized in this work. Thus, the study should be of value to the kind of research proposed here, but the orientation is not primarily to resources.

The Western European forest products export market. The analysis has shown that Western Europe is and will probably continue to be the primary source of external demand on the timber resource of North America. Thus, the Western European forest products markets would apear to be a logical area for continuing research related to the level of demands on North American timber which should be anticipated.

However, it should be noted that a large volume of material related to this area is already available in the United States. The analyses of the FAO, the various reports and studies of the Timber Committee of the Economic Commission for Europe, and such studies as the current work on pulp and paper being conducted at the University of Stockholm are examples of this.

The primary current need seems to be for detailed market analyses by specific types and grades of forest products designed to identify favorable market opportunities and means for market development. In general, such studies fall primarily within the province of industrial marketing studies appropriate to individual companies, either directly or through consulting organizations.

The timber resource relationships of Japan. Japan presents a particularly promising opportunity for research which can contribute both to an understanding of the relationship between economic development and resources, and to guidance for policy decisions affecting the timber resources of the United States.

As the preceding analysis has shown, Japan is the focal point of timber trade in the western Pacific. The economic development of Japan has proceeded at a remarkable rate, yet the natural resources of the nation appear very limited relative to its population. In the case of timber, this has led to a far-flung system of wood imports through which Japan has become a significant force in the market for logs in the Pacific Northwest, in the development of Alaska, in the development of eastern Siberia, in the demand for logs from the Philippines, North Borneo, and other tropical hardwood sources, and in the market for lumber from British Columbia. It is also the leading source of the hardwood plywood and veneer exports to North America, although its production in this field is based largely on imported logs. The possibility exists that in time it will become an important bridge to Mainland China.

Despite the level and intensity of this timber-related activity, no extensive research studies of the resource relationships involved have been undertaken. Thus, although the total demands on North American timber resources are considerably less than from Western Europe, the opportunity for significant research appears to be far greater in this instance.

The role of the U.S.S.R. From the standpoint of North American timber resources, the U.S.S.R. is of interest primarily as the main alternate source of softwood forest products in the world export markets. Greater knowledge of this potential and its probable rate of development would be very helpful in planning the development of the North American timber resource.

The greatest immediate need in this area appears to be a form of economic intelligence. A substantial amount of information concerning forestry and the forest products industries of the Soviet Union is potentially available, but an organized and sustained effort is required to realize this potential. Significant information is published in a wide range of scattered sources, including a substantial number of standard statistical publications; occasional articles in both the periodical and daily press; at least fifteen forestry, forest products, and related journals; and numerous special pamphlets and books. The job of translating, collating, and interpreting these scattered materials is a large one, requiring organization on a sustained basis. As far as can be determined, there is now no regular source for such a program for general economic purposes. The U.S.S.R. is not a member of the FAO, and that organization has no provision for a Russian expert in its forestry and forest products division. Some occasional work of this kind at various universities indicates that an effective interpretive job could be done if a regular provision were made for the work.[1]

The comparative costs of expansions in forest products output. In attempts to determine the possible resource impact of projected increases in forest products consumption levels in various parts of the world, the essential question to be faced is the ex-

[1] A notable example of the possibilities of such research is provided by the detailed analysis of forestry and timber development in the U.S.S.R. by Dr. K. V. Algvere of the Department of Forest Economics, College of Forestry, Stockholm, Sweden, entitled "Forest Economy in the U.S.S.R.—An Analysis of Soviet Competitive Potentialities," in *Studia Forestalia Suecia* Nr. 39, 1966. Another study of the forests of the U.S.S.R. is reported to be in the process of publication by Dr. Ivan M. Elchibegoff of the United States.

tent to which expansions of outputs in North America will be a more economical means of serving these markets than expansions in other regions of the world. As has been discussed in the preceding analysis, the simple gap model used by the FAO in its studies of timber trends does not give adequate recognition to this question. Basically the FAO estimates the import component by regions as the gap between the projected consumption level and the regional output level estimated as possible from a silviculturally determined allowable cut. Actually, imports may become a cheaper means of supplying markets than domestic production long before such allowable cut limits are reached. Although protectionist policies may reduce the significance of this question in a single nation, it emerges with particular importance when the issue involves the role of two or more exporting regions relative to a separate importing region. For example, during the course of this study a question repeatedly encountered in Northern Europe was that of the relative cost of expanded output of pulp and paper from North America as compared to the cost of expanded output from the Northern European nations. The focus of the question was the extent to which Northern European producers should expand output in the face of rising costs in the expectation of serving expanded Western European markets.

If research is to contribute to the formulation of appropriate growth goals and product development goals for North America as a net exporting region in the world trade in wood, then information must be developed as to the comparative costs of expanding forest products output in North America and in other major forest regions of the world. Only the most fragmentary information now exists on such cost structures, and current judgments in this field tend to be largely subjective impressions with a minimum of underlying analysis.

Requirements for a resource inventory of timber. Timber inventories conceived as a physical stock-taking of forests are in progress in many parts of the world. Resource inventories of timber, conceived with full recognition of the functional nature of a resource, have not yet been initiated. A very large amount of research has been directed to methods of sampling, problems of measurement, and the use of aerial photographic techniques in the conduct of timber inventories. Very little organized research has been directed to the question of the information which can and should be collected to give the timber inventory an economic content relevant to the functional dimension of the resource.

In the course of this study the question was repeatedly encountered as to the extent to which the forests of Canada and the U.S.S.R. were actually a functional resource. The question would be phrased in terms of rivers flowing to the north, or of rates of growth possible in the permafrost zone, or of remoteness from the main consuming centers, but the central issue was always that of the potential function of the timber. In timber inventories, the question is often implicitly recognized through classification of forests as "accessible" or "inaccessible." Such classifications are usually based on some physical criterion such as distance from an established transportation route. As has been pointed out earlier in this report, in an economic sense timber in close proximity to transportation is equally inaccessible when its characteristics are such that it cannot be harvested and transported at a cost that can be covered by the value of the resultant product.

The need is to define a pattern of inventory data appropriate to economic analysis. Alternatively phrased, the need is to develop a measure of the supply of timber, using the word "supply" in its pure economic sense, rather than simply the physical stock of timber. The problem is a conceptual one, but it has distinct applied phases since the data requirements must be specified in terms which can, in fact, be developed through the inventory program.

The approach might involve a rigorous analysis of the conceptual requirements for a set of measures defining a resource in its functional sense combined with critical review of the various approaches which have been attempted in making inventories both of timber and of various other resources. The outcome of such an effort is highly uncertain, for the measurement problem involved may well be most intractable and not susceptible to definitive solution. However, both the importance of developing data more suitable to economic analysis, and the limited likelihood that the agencies responsible for timber inventories will attack the problem are such as to justify research in this area.

The economics of forest resource development. Although this study has focused on the relationship of world developments to the forest resources of the United States, it is not possible to carry out such an inquiry without developing an intensified awareness of the problems of resources and economic growth in the extensive underdeveloped regions of the world.

The view has been advanced that a resource has significance—indeed, that a resource *is*—only in terms of technologies, demands, and complementary resources. While progress is being made in the development of forest inventories and the determination of areas of productive forest land in these underdeveloped areas, the realization of these physical stocks as meaningful resources remains generally remote. The United States (along with various other developed nations) is involved in the processes of resource development in forestry on many fronts—through AID programs, through Peace Corps contributions, through consulting firms, through industrialization under the aegis of U.S. corporations, through university programs, and through strong financial support of the FAO programs (though its role here is not always recognized).

The problem of forest development must be recognized as one involving a fully integrated relationship between the resource and the resource-based industry within some total economic environment. The forest area and its timber must be developed and managed in balance with the industrial structure using the wood as its raw material. Such a balance must be conceived and planned over a long period of time, with shifts in the nature of the industrial structure being phased and matched to shifts in the nature of the raw material as the forests are moved through successive stages of exploitation and development.

In the various efforts now under way, there exists a tremendous gap between the initial fact-finding surveys and the realization of effective forest development. Personnel competent in planning and following through in this developmental stage are scarce, while demand for such services is rising rapidly. The problem is made the greater by the fact that there exist today no effective programs for educating individuals for such work. Indeed, the material which would be required for such an educational program has not been brought together in organized form.

There is thus an increasingly urgent need for a thorough study of the economics of integrated forest development, covering both the theoretical and the institutional aspects. North America has had a rich experience in such developmental work and thus provides a fruitful opportunity for economic analysis of the processes involved. And it might be added that, despite this rich experience, the means of achieving and stimulating such development are not yet fully understood, as is witnessed by the current problems relative to Appalachia and similar forest regions in the United States.

In any such study, attention must be directed to the historical, political, and economic context of the experience. This context is obviously very different for the less-developed economies at the present stage in history than for North America. Unless the significance of these forces is fully recognized, any attempt to abstract from the North American experience could be seriously misleading. Thus, while such a study might draw heavily on the North American experience, it should also involve comparative analyses of some of the European experience and some of the current developmental work in the less-developed regions.

Case studies of forest resource development under alternative public land policies. Within the United States 47 per cent of the total volume of growing stock on commercial forest land was held in public ownership as of January 1, 1963. These public timber holdings were largely concentrated in the West. The U.S. Forest Service alone administered 79 per cent of this public timber volume. The policies under which the public timber is administered are a major factor in the determination of its function as a resource and of the kind of resource development which takes place. With the dominant role of a single agency in such public timber administration in the United States, the opportunities for the analysis of possible alternative policies have been limited.

In the developing economies of the world, public ownership of timber and land is widespread. Decisions must be made as to the policies on these public lands appropriate to resource development.

A series of case studies analyzing resource development under alternative public land policies could be of substantial value in the development of timber resource policies on public lands both within the United States and in the various developing nations.

The variety of timber resource development programs for public lands to be found among the nations of the world provides an opportunity for a range of comparative case studies which is almost entirely lacking within the United States. For example, British Columbia is following a program of long-term management licenses designed to stimulate economic development of the province. This policy differs fundamentally from the U.S. Forest Service emphasis on short-term sales and a maximization of competition, with the sales being developed in response to economic activity rather than as a means of stimulating economic growth. In view of the substantial general similarities which exist between the Pacific Northwest and British Columbia, this provides an unexcelled opportunity for a comparative case analysis.

Numerous other examples could be developed. Switzerland provides a case involving primary reliance on public ownership of forest land, but with that ownership being almost entirely at the local level. The result is a pattern of public ownership in which holdings are small and management decisions are completely decentralized, providing a sharp contrast with the highly centralized co-ordination of decision-making characteristic of the public timber resource of the United States. In Sweden, the public timber agencies are moving strongly into forest products manufacturing, providing an approach basically different from the traditional U.S. public timber policy of not even engaging directly in the harvesting operations.

Such a series of case studies could be a very effective basis for the development of an internationally oriented research program in timber resources, with results having importance within the United States as well as in other areas of the world.

REFERENCES

[1] British Columbia Bureau of Economics and Statistics. *Recent Developments in the British Columbia Pulp and Paper Industry.* Victoria, British Columbia, 1964.

[2] Canadian Department of Forestry, *Summary. Forest Inventory Statistics, Canada, 1963.* Ottawa, Canada, 1964.

[3] Dominion Bureau of Statistics. *Canadian Forestry Statistics, 1961.* Ottawa, Canada, 1962.

[4] Duerr, William A. *Fundamentals of Forestry Economics.* New York: McGraw-Hill Book Co., Inc., 1960.

[5] Fisher, W. Halder. *Analysis of the Relationship of Softwood Log Exports to the Economy of the State of Washington.* Columbus, Ohio: Battelle Memorial Institute, 1964.

[6] Food and Agriculture Organization of the United Nations. *Timber Trends and Prospects in the Asia-Pacific Region.* Geneva, Switzerland, 1961.

[7] ———. *Latin America Timber Trends and Prospects.* New York, 1963.

[8] ———. *Pulp and Paper Prospects in Latin America.* New York, 1963.

[9] ———. *Yearbook of Forest Products Statistics, 1964.* Rome, Italy, 1964.

[10] ———. *European Timber Trends and Prospects: A New Appraisal, 1950–1975.* Rome, Italy, 1964.

[11] ———. "The Implications of Mainland China's Situation for Regional Timber Trends and Prospects." 1964 (mimeo).

[12] ———. *Prospects for Expanding Forest Products Exports from Developing Countries.* Rome, Italy, 1964.

[13] ———. "Fast-Growing Tree Species for Industrial Plantations in Developing Countries." *Unasylva,* 19(4):159–67.

[14] ———. Advisory Committee on Pulp and Paper. "Pulp and Paper Trends in the U.S.S.R. and Eastern Europe." 6th Sess., Doc. No. 12. Rome, Italy, 1965.

[15] ________. "Wood: World Trends and Prospects." *Unasylva*, 20(1–2):1–133.

[16] ________. *World Forest Inventory–1963*. Rome, Italy, 1966.

[17] Forestry Agency of Japan. "Forestry in Japan." Tokyo, Japan, n.d.

[18] Love, D. V. "Potentialities of the Forest Resource Base," in Resources for Tomorrow, *Conference Background Papers*, Vol. II, 1961, pp. 641–60.

[19] Masson, F., and H. E. English. "Invisible Trade Barriers Between Canada and the United States." Washington, D.C. and Montreal, Canada: Canadian-American Committee, 1963.

[20] Richardson, S. D. *Forestry in Communist China*. Baltimore: The Johns Hopkins Press, 1966.

[21] Solecki, Jan J. "Forest Resources and Their Utilization in Communist China," *Forestry Chronicle*, 40(2):227–46.

[22] ________. "Competition from Russia," in T. Matthews (ed.). *British Columbia's Future in Forest Products Trade in Asia and the Pacific Area*, Proceedings of Seminar, University of British Columbia, February 19, 1965, pp. 67–82.

[23] U.S. Department of Commerce, Far Eastern Division, Bureau of International Commerce. "The Japanese Market for Softwood Logs and Lumber." 1964 (mimeo).

[24] U.S. Forest Service. *Timber Depletion, Lumber Prices, Lumber Exports, and Concentration of Timber Ownership*. Report on Senate Res. 311, 66th Cong., 2nd Sess. Washington, D.C.: U.S. Government Printing Office, 1920.

[25] ________. *Forests and National Prosperity*. U.S. Department of Agriculture Misc. Pub. No. 668. Washington, D.C.: U.S. Government Printing Office, 1948.

[26] ________. *Timber Resources for America's Future*. Forest Resource Report No. 14. Washington, D.C.: U.S. Government Printing Office, 1958.

[27] ________. *The Demand and Price Situation for Forest Products, 1964*. U.S. Department of Agriculture Misc. Pub. No. 983. Washington, D.C.: U.S. Government Printing Office, 1964.

[28] ________. *Timber Trends in the United States*. Forest Resource Report No. 17. Washington, D.C.: U.S. Government Printing Office, 1965.

[29] U.S. Tariff Commission. *Softwood Lumber*. Tariff Commission Pub. 79, Washington, D.C., 1963.

[30] Vassiliev, P. V. "U.S.S.R. Forest Resources and Features of Their Inventory," *Unasylva*, 15(3):119–24.

[31] Zimmerman, Erich W. *World Resources and Industries*. New York: Harper and Bros., 1951.

[32] Zivnuska, John A. "The 1964 Timber Trends Study in Perspective." *Proceedings*, Society of American Foresters, 1964, pp. 84–88.

GLOSSARY

As do specialists in other fields, those who concern themselves with the study of timber and the timber industries have a specialized language. Since it is hoped that this study will be of interest to readers other than students of forestry, some of the field's terminology has been defined below.

Allowable cut. The volume of timber that may be cut during a given period under specified management plans for sustained production.

Cooperage. Containers consisting of two round heads and a body composed of staves held together with hoops.

Cull trees. Live trees of sawtimber and poletimber size that are unmerchantable now or prospectively because of roughness, rot, or species.

Fiberboard. Sheet materials of widely varying densities made of refined or partially refined wood (or other vegetable) fibers with or without addition of bonding agents or other materials to increase strength, resistance to moisture, fire, or decay, or to improve some other property.

Fuelwood. Wood to be used as fuel for purposes of cooking, heating, production of power, etc. Includes wood for charcoal pit kilns and portable ovens and may include wood from trunks and branches.

Growing-stock trees. Live sawtimber trees, poletimber trees, saplings, and seedlings meeting specified standards of quality or vigor; excludes cull trees.

Hardwoods. Dicotyledenous trees, usually broadleaved and deciduous. All woods derived from trees classified botanically as

Angiospermae—e.g., maple, alder, ebony, beech, lignum vitae, poplar, oak, teak, etc.

Industrial wood. All commercial roundwood products except fuelwood; includes the following roundwood: sawlogs, veneer logs, logs for sleepers, pulpwood, pitprops, other industrial wood.

Newsprint. Bleached, unsized or slack-sized printing paper, without coating, of the type usually used for newspaper.

Net annual growth of growing stock. The annual change in volume of sound wood in live sawtimber and poletimber trees resulting from natural causes.

Paperboard. The term usually designates all grades of fibrous material which are 0.012 inch or more in thickness made on a paperboard machine. Some examples of paperboard are: pulp boards, felt boards, duplex, triplex, and multiplex boards, parchment paperboard, composite paperboard, corrugated paperboard, embossed paperboard, coated or impregnated paperboard, etc.

Plywood. Sheets of wood material made up from three or more layers of wood glued or pressed one on the other and generally disposed so that the grains of successive layers are at an angle.

Poletimber trees. Live trees of commercial species at least 5.0 inches in diameter at breast height but smaller than sawtimber size, and of good form and vigor.

Pulpwood. Coniferous and broadleaved wood and chips for pulping by mechanical (including defibrating, exploding, etc.), chemical or combined means. In some contexts this classification may include wood residues except sawdust.

Roundwood. Logs, bolts, or other round sections cut from trees.

Sawtimber trees. Live trees of commercial species containing at least one saw log. Softwoods must be at least 9.0 inches in diameter at breast height, except in California, Oregon, Washington, and coastal Alaska where the minimum diameter is 11.0 inches. Hardwoods must be at least 11.0 inches in diameter in all states.

Softwoods. Coniferous trees, usually evergreen, having needle or scalelike leaves. All woods derived from trees classified botanically as *Gymnospermae*—e.g., fir, Paraná pine, deodar, ginkgo, larch, spruce, pine, chir, kail, etc.

Sustained yield. As applied to a policy, method, or plan of forest management, implies continuous production with the aim of achieving, at the earliest practicable time, an approximate balance between net growth and harvest, either by annual or somewhat longer periods.

Veneers. Thin sheets of wood of uniform thickness, rotary cut, sliced, or sawn, for use in plywood, laminated construction, furniture, veneer containers, etc.

Volume of growing stock. Net volume in cubic feet of live sawtimber and poletimber trees from stump to a minimum 4-inch top (of central stem) outside bark or to the point where the central stem breaks into limbs.

Wood pulp. Fibrous material prepared from wood by mechanical and/or chemical processes for further manufacture into paper, paperboard, fiberboard, or other cellulose products.

Designed by Gerald A. Valerio
Composed in Caledonia by Baltimore Type
Printed offset by Murray Printing Company, on P & S Olde Forge, F
Bound by Murray Printing Company

For Product Safety Concerns and Information please contact our EU representative GPSR@taylorandfrancis.com
Taylor & Francis Verlag GmbH, Kaufingerstraße 24, 80331 München, Germany

www.ingramcontent.com/pod-product-compliance
Ingram Content Group UK Ltd.
Pitfield, Milton Keynes, MK11 3LW, UK
UKHW022157190926
475702UK00003B/73
* 9 7 8 1 1 3 8 8 5 7 2 3 0 *